# of Pork & Potatoes

– a memoir –

BILL MASSEY

Bill Massey

Suite 300 - 990 Fort St
Victoria, BC, V8V 3K2
Canada

www.friesenpress.com

Editors, Jenny Gates and Phyllis Braun
Cover illustration by @lucillustrations

First Edition — 2021

ISBN
978-1-5255-8554-8 (Hardcover)
978-1-5255-8553-1 (Paperback)
978-1-5255-8555-5 (eBook)

1. Biography & Autobiography, Personal Memoirs

Distributed to the trade by The Ingram Book Company

TABLE OF CONTENTS

INTRODUCTION

HELLO, THERE! I'M BILL MASSEY, AND I live on our little farm at Big Island with my wife and partner, Dorothy. I'm a maverick. I didn't start out to be a maverick. I didn't even want to be a maverick. So how did this happen?

I'm seventy-one years old and a retired school principal. I should be kicking back and enjoying these golden years, but here I am, involved in the battle of a lifetime.

For some reason, I chose a different path. I started out wanting to farm and to teach, and I've done both for many years very happily. I have principles. I *was* a principal for twenty-three years, for heaven's sake! I loved it and I was good at it.

One thing I don't love is injustice. In fact, I hate injustice. I hate it so much I can feel it in my joints and in the pit of my stomach. I stand up for the rights of the less fortunate. I also don't like bullies much, and I take them down whenever I have the opportunity. I strive to walk the high road in all things. I like leading. I like leading people in the accomplishment of amazing goals and seeing them flourish in reaching their objectives. I ask no more of anyone than I expect from myself. So how did I get to this point?

My wife Dorothy and I both grew up on small farms in the 1950s and '60s. Our families each had a quarter section of land, and our dads supplemented their incomes by working off the farm. Dorothy loved the little farm where she grew up. She and her siblings had names for all the little bluffs around the sloughs. Out in the fields, they played

Robin Hood and other imaginary games. Dorothy's dad was a carpenter, and my dad was a mechanic. Dorothy's family raised chickens, while we had pigs.

These days we raise chickens and pigs on our small farm, and also have a small flock of sheep. Dorothy has a big garden in the summertime. She makes wool blankets and does other fibre arts. We live much as we did fifty years ago. We raise food for our children and grandchildren and have the tastiest and healthiest meat and vegetables you could possibly imagine. We enjoy our community and are very involved in volunteering in a variety of ways. We have made many friends over the years and try to help anyone we can.

Dorothy was also a teacher, and since we retired, we have both done a variety of part-time jobs, as well as farming. I have particularly enjoyed working for several companies restoring original tall grass prairie. We love looking after our four grandchildren whenever we have the opportunity. They happily and excitedly come here to experience the warm summer breezes, the farm animals, and fun winter activities in the snow. We're very fortunate that they all live close by and that we can see them often. We have a great life.

This book blends my childhood in rural Manitoba with our community's fight against the excesses of the hog industry, as well as my role of leadership in this struggle. When I became a teacher and then a principal, I had the opportunity to be an advocate and did a lot of work with abused children and children in poverty. I understood them because I was one of them. It was not a big step for me to move from advocating for disadvantaged children to dealing with the injustice I saw in my community. So why did I decide to do this? Because my integrity would have not let me sleep otherwise.

Thank you for choosing to read this book. It describes a struggle that I've been engaged in now for the past fifteen years, and another that I was part of more than fifty years ago. It is a story of a journey to thrive and flourish, and the will to overcome adversity. You will know me much better by the time you are finished. Come walk this road with me.

CHAPTER ONE

Is It Not Enough?

"IT'S FOR YOU," DOROTHY CALLED. "IT'S THE colony!" Happily, I picked up the phone. I enjoyed associating with the colony because I liked the people and I'm fascinated with the culture.

"I got a load of wood for you," Jacob said in his charming Hutterite drawl. "And don't forget the paper."

Dorothy and I burned wood to help heat our house, and Jacob was gathering some up at the colony for me. "Should I bring the big truck or the little truck?" I asked.

"Well, you better bring the big truck this time, and don't forget the paper!"

Jacob and I had an arrangement. If there was a lot of wood, I could bring the big truck, and if there wasn't very much, a pickup would do. Jacob especially liked getting the paper because when there was a big load, there would be a twenty-six ouncer rolled up in the paper. For a small load, there would be a Mickey. It was a good arrangement for both of us.

The colony was a busy place. Farm trucks were always coming and going. Many people from our community had equipment repaired or built there because they were reasonably priced and did a great job. Jacob and I had a prearranged place to meet. Getting the wood was no problem; the newspaper and its contents, on the other hand, were a different matter—Jacob's brothers were the colony's religious leader, their business manager, and the farm boss. The three brothers were in charge of the colony and aware of most of the transactions that went on. There is quite the hierarchy on the colony, and Jacob did not seem to be at or even near the top of the heap. That was one of the reasons I liked him. I like people like that; there is a side of me that easily relates to them. Nothing cliquish about Jacob.

After I handed over the newspaper and its contents, Jacob got into the truck and we headed over to load the wood. We spent half an hour loading, and then nothing would do but that I had to go in for coffee. Jacob's wife, Mary, had also just baked some delicious rolls. Jacob told me about the trip he had made to the U.S. where he had visited some Amish people. He started imitating their style of speech. I found it so delightful to hear an Amish dialect spoken with the Hutterite accent. I told him I needed a large piece of cement moved on my farm. It was the floor from the old shop. I wanted to put it in front of the new building and I didn't have anything of my own big enough for the job. Jacob said he could bring the Huff loader over and move it for me for a hundred dollars. I thanked him and Mary for their hospitality, and we arranged to move the cement the following week. I got in my truck and headed home. I enjoyed my visits to the colony.

The farm Dorothy and I have is just northwest of Big Island, Manitoba. We own twelve acres, and our neighbour lets us use another five acres, which is locked in by bush next to our place. When the province moved Highway #6, we got the rights to hay another five acres. Two other neighbours let us make hay on two small fields of about eight acres in total. Altogether, we farm nearly thirty acres. It's small, but big enough so I can do everything I want in farming.

Dorothy was good friends with Jacob's three single sisters in the colony; the oldest one even taught her to spin wool. When Dorothy was young, her family enjoyed spending time with people at the colony nearest their home. She remembers great times and good friendships and was looking forward to developing those relationships with the people at this colony. As for myself, I repair and sell small farm tractors and equipment suitable for acreages. We both love life on our little farm. We have neighbours all around us, including the colony, which is across Highway #6 and just under a mile away. The Village of Big Island is about the same distance southeast of us. We have a beautiful place.

***

IT WAS JUNE 2004. I DROVE INTO Big Island to pick up the mail. It was a cloudy, cool day with the north breeze blowing. I wished I had worn a jacket. That day, there was an invitation in the mail. It was from the colony to attend a meeting to discuss the expansion of their farm. It included a letter from DGH Engineering indicating that they had been retained by the colony to assist them in developing a plan for the project. We had received this invitation because we were within a mile of the proposal. I talked with a few of the neighbours, and found out that the colony was proposing an expansion of their hog facility. I got a sinking feeling in the pit of my stomach when I heard that news. Some of Dorothy's family lives in southeastern Manitoba, an area saturated with hog barns, so we all knew what this could mean for us.

On the day of the meeting, Dorothy and I finished work in the garden—cleaning out the first weeds of summer—and headed over to the colony. There was hardly room to park. Most of the community was there, as well as DGH Engineering, the Pork Council, feed companies, and others.

Representatives from DGH led the meeting, and according to them, their plan was amazing. It considered environmental concerns and provided economic development for the rural municipality and our community. It was a slick, polished, professional presentation that obviously had been presented many times before. I had often seen those types of presentations, and recognized it for what it was. A snow job. It seemed a little more sincere when Peter, the business leader of the colony, got up and expressed the colony's desire to work closely with the community in the development of this project. Promises were made that every concern would be considered and dealt with. It seemed to me there was almost an air of apology on the part of the colony. I wondered if perhaps not everyone in the colony was in favour of this project.

Then the questions began. Big Island sits on a vast underground aquifer. How was the colony planning to keep the aquifer from being polluted? There were several uncapped wells from old homesteads near both the manure storage facility that was already in place and the proposed location for the barn. Those wells had never been properly sealed. What about them? People who had lived in this area their whole lives knew about the wells. There was an outcropping of limestone where a small limestone factory had operated within half a mile of the proposed site for the barn. The aquifer was fed from the surface through openings in the limestone. Could polluted water gain access there? Odour was another big concern for the community. There was a school in Big Island, and a number of the neighbours were raising small children. One of the people present worked in public health and expressed their concerns about air pollution and airborne illnesses. Also on the list were increased road traffic and decreased property values.

Then the subject of surface water pollution was brought up. The existing hog facility was already leaking manure directly into Colony Creek, which flows into Omand's Creek, and eventually into the Assiniboine River. Jacob stood up and took responsibility on behalf of the colony for that environmental spill that was occurring. We were all impressed with his boldness in speaking out about this problem. Several of us speculated that he was probably going to be disciplined—shunned—after his remarks at the meeting. The colony leaders' response was that all of these issues would be taken care of with the new development. We wondered why they hadn't been dealt with already if they were aware of the problems.

A number of community members who'd had dealings with the colony spoke eloquently in expressing their concerns. I listened carefully, but said nothing. Once again, I had this distinct feeling in the pit of my stomach, and my joints—the kind I get when I experience impending injustice. I could sense that a storm was brewing.

Dorothy and I drove home in silence. I wanted to believe the colony's assurances that this development would not have a negative impact on the community. But the more I considered it, the more I realized that there were just too many red flags. I was afraid that this big corporate entity was about to do a number on our community. I knew I could not rest if I didn't do something. There was the smell of bullying in the air.

The next day, I went to see Harold. Harold was a community leader and affectionately called the mayor of Big Island. We talked about our concerns, and we were of the same mind. We were both uneasy about the effect on the community and our quality of life. I talked to several other neighbours, and all of them were quite worried about what was happening. We decided to call a meeting.

***

JUNE 6 WAS IN THE MIDST OF the growing season in Manitoba. The air was fresh and the smells were beautiful. Lovely, warm breezes drifted through the yard in the sunshine. Twenty-one of us gathered in my neighbour's basement. We talked about our concerns, including pollution of the aquifer, surface water, air pollution and odours, health concerns, road traffic, and decreased property values. A number of us agreed to carry out specific tasks. One person had contacts with the University of Manitoba's Faculty of Agriculture, and we were interested to hear what an academic had to say about this situation. Another person would contact the Friends of Sturgeon Creek, an environmental group concerned with the health of Sturgeon Creek; the proposed hog barn was located in the Creek's watershed. Others had contacts with Hog Watch Manitoba, a group promoting a sustainable hog industry in Manitoba. Someone would contact Ducks Unlimited Canada, an international organization involved with Grant's Lake Wildlife Management Area less than a mile from the manure storage facility. I would contact the Province's Department of Natural Resources. Someone else would contact the local public health regional office and Interlake School Division. Another was going to talk to the local municipal governments. I was also to contact the new government department of Water Stewardship and Biodiversity, and research Bill 40, livestock manure and mortality, and management regulations.

The final business for that evening was to select a leader. While there was considerable joking about this, the group put considerable pressure on me to take on that role. I was reluctant because I knew what it would do to my relationship with the colony. I also knew there was a lot to learn to be able to deal with this matter. I had retired the year before and felt I had done a good job of leaving that life behind. I just wanted to immerse myself in farming and let everything else go, but my concerns about the dangers that seemed to be looming over our community would not let me.

Dorothy also had friends at the colony, and she was worried how that would impact her friendships now as well as in the future. I looked to where she was sitting across the room, and she privately shook her head. I could tell that she did not like this at all and really did not want me to take on the leadership of the group.

Dorothy and I talked that evening after the meeting. She talked about the conflict and difficulties I had been exposed to growing up. She went on to say my job, especially as a principal, had been challenging, and I had often faced controversy and opposition. The school division had given me difficult situations, students, and teachers to deal with because they believed I could do the hard jobs for them. I had also faced some personal challenges in my life, and with Dorothy's help, I had come out the other side relatively intact. These struggles had been hard on her as well, and through it all, she had always done her best to support me. Finally, she asked, "Is it not enough?" I think she knew the answer to that question before she even asked it.

Harold came to see me and told me I was probably the best person to take on this task. He really leaned on me, and I did not want to let him down. I had worked closely with Harold on many of his community projects. I felt that the two of us were a team, and this felt like another project we were doing together for Big Island. I sat down with Dorothy and talked quietly at length about my concerns if I didn't step up, and although she did not like it at all, she gently said, "Do it if you must." She understands me.

I pondered for several days until I realized I could not refuse.

Nevertheless, I felt troubled because I knew how this would likely end up. I called Jacob and asked if I could come over and see him. As I drove to the colony, I considered how I was going to tell him. We talked for a while about the meeting and the proposal, and he asked me what I thought. I told him I was impressed with his honesty and hoped he hadn't gotten into trouble with the colony leaders. Then I said, "I have something to tell you. I was asked to lead the group that is going to oppose expansion of the hog barn, and I have accepted." I could see his face tense up as he listened to the information. He became very quiet.

"Let's go in for coffee then," he suggested, and we walked silently into the house. Mary had the coffee made and was putting fresh bread on the table. Jacob turned to her and said, "Bill is leading the group that is going to oppose the hog barn."

"Well!" she exclaimed. "Well, let's talk about something else."

She took me over to the window to see her flower garden just outside. We admired the daffodils, irises, and roses, which were flourishing in a myriad of beautiful colours. We had coffee and broke bread together. And then I got up, wished them well, and left.

I have not seen my friends for fifteen years. I think about them often. But things would never be the same again.

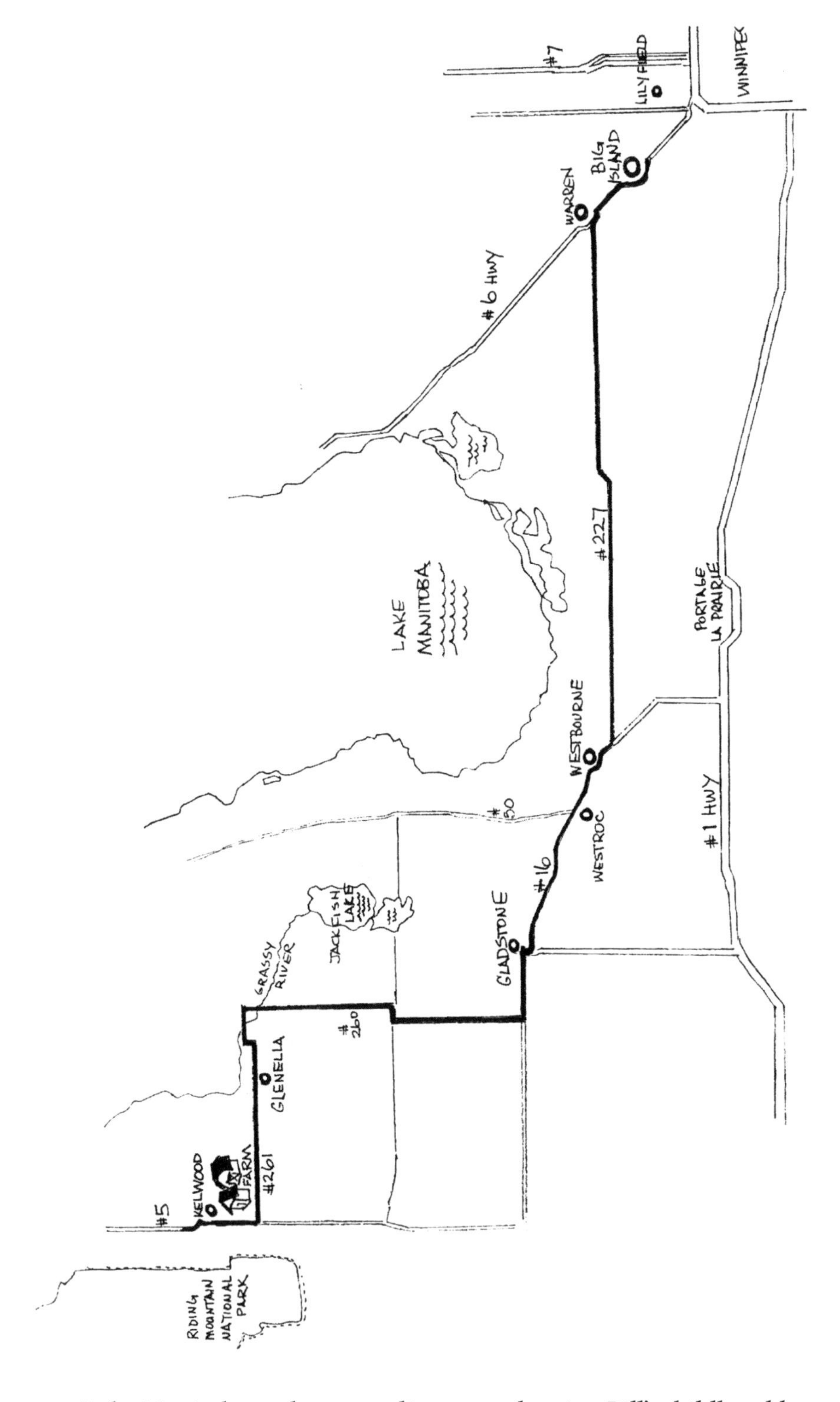

Lake Manitoba and surrounding area, showing Bill's childhood home.
Drawn by local artist Linda Gillies.

CHAPTER TWO

The Beginnings

I REMEMBER MY MOTHER CRYING A LOT. My father worked as a mechanic for the army at Shiloh. He had served in the Second World War, but was a civilian now. We all lived in the barracks on the base, and several of Father's single friends lived in the barracks down the hall.

The story goes that my sister Marlene would crawl down the hall with a diaper in her mouth to get changed. I was only a babe in arms at the time, and have always wondered why that happened. Did Mother not look after her, so the friends had to intervene? I knew those men later when I was a young adult, and they were good people. None of them ever talked to me about it.

One of my first memories is pulling my sister in my little red wagon. It was 1951, and I was two years old. Marlene was a year older than me, and had come down with polio. That would be the reason for my mother's tears, and the reason I was pulling my sister around. As far as polio victims were concerned, Marlene was one of the luckier ones. She recovered the use of her legs, but required an operation on her right foot when she was about twelve years old. I remember my parents describing how in the midst of the polio scare, she had suddenly been unable to walk or stand up. I always thought my dad favoured her, but that would soon change. I have always wondered how that childhood illness changed her, and what she would have been like if she had not been sick at an early age.

The second vivid memory I have at that age is my paternal grandmother. We went to see her the summer I was three years old, and I remember this large woman sitting in a rocking chair, with a bucket beside her that she used as I watched. When she glanced up at me from that bucket, there was such pain in her eyes, not only as a result of the pain of the breast cancer that was killing her, but much, much more. I realize now, as a parent and grandparent, that it was the pain of knowing that she was not going to be around to care for me and love me. There were so many things that she could see about my family and the relationship between my parents that probably caused her great concern. My memory of that glance is as clear and as real today as when it happened sixty-seven years ago. She is a mystical person in my life and I sometimes wonder if she isn't near.

I have a memory of us living for one summer in Wasagaming, the town at Clear Lake, Manitoba. My dad got a job as a mechanic in a garage. One day he got a call that there was a problem on the lake. He called me to come along and we went out in a rowboat to a cruise boat that had stopped on the lake. The cruise boat had an old car engine that had quit running and left the passengers stranded. Dad climbed into the boat and, within fifteen minutes, had the engine running again. As we left to row back to shore, the passengers in the boat clapped and cheered, which embarrassed him, but I felt very proud and believed he could do anything. Dad was a wonderful mechanic and a great trouble shooter.

We lived in one of the little cabins in Wasagaming. We would go with my mother to the nearest park and feed peanuts to the squirrels. One time I teased the squirrels by holding out my fingers with no nuts in them, and I got bit for my troubles. Then my sister started to tease me, so I bit her, and she ran crying to Mother. Mother had seen the whole thing and was not very sympathetic. After that, whenever Marlene wanted to tease me, all I had to do was show my teeth. I think I learned something about adversity that day.

Because Wasagaming was a tourist town, it shut down for the winter, so we moved to a house in Neepawa to spend the winter. My dad had also bought a farm at Kelwood, but didn't take possession until the spring. Veterans Affairs had lent him the money at a very low interest rate; they always treated him very well.

The memory I have of our time in Neepawa is getting my dog, Sport. He was a black and white puppy, and cute as a button. One day in March, I decided he needed a bath. There was a tub outside filled with water from the snow melting on the roof. I grabbed little Sport and in he went. I gave him a good washing and got soaking wet myself, so I ran inside to change into dry clothes. Sport was an outside dog because my mother didn't want animals in the house.

When Mother saw Sport, she cried, "What have you done to that poor puppy?"

"Well, I thought he was dirty and needed a bath!" I answered. I have never again seen an animal shiver like that.

Nothing would do except to bring that pup inside, and I dressed him in my pyjamas. I still have this image in my mind of Sport running around the house with pyjama legs dragging out behind him. My dog and I became inseparable.

***

WE MOVED TO THE KELWOOD FARM IN the spring of 1954. I had found my heaven. There were two roads into the yard, and the best one was through the bush. It was late March, so there was still eight feet of snow on that road. The second road was the trail across the field, probably the lowest part of the farm. I remember Dad spinning across that field through the mud in our old 1940 Chevy car. Soon after we moved in, we bought a team of horses. Marlene and I were fans of Dale Evans and Roy Rogers, so the horses were called Trigger and Buttermilk. As it turned out, those first few years were quite wet, and our horses were essential for getting on and off the farm. I was so impressed with those big strong animals and what they could do, and I loved their horsey outdoor smell.

Dad bought a 1938 John Deere Model G Unstyled row crop tractor in Neepawa, as well as a high-wheeled wagon with a grain box. He put me in the grain box, and we headed out to Kelwood with our tractor. I was five years old. I could stand in the grain box and just see over the front. We were doing six-and-a-half miles an hour on the rough gravel road, top speed for the old tractor. It was about a five-hour trip, a long time for a little boy to stand and be shaken up and down. But I was game and never complained, although I was pretty tired by the end of it. That was one of the first endurance tests in which I would find myself.

That first summer on the farm was wonderful. Sport and I played outside all the time and we came to know that bush, most of the trees, and every dip and hollow. It was a five-acre paradise for a boy and his dog. Mostly there was poplar, but also some oak, elm, and willow. On the west side was a blow dirt bank where airborne soil had collected during the drought in the thirties. I was so excited to be there and outside, my mother could hardly get me inside to eat.

In September, Marlene headed off to Glenallan School to attend Grade 1. We made friends with other children in the district, and we went to picnics and birthday parties. Sport and I stayed home that winter, and he and I got to know the farm and the snowbanks inside and out. Every snowbank had to be examined for potential tunnels and forts. The two of us were always together.

One time, I got into trouble with my mother, and she was going to spank me. I managed to get away from her and ran outside. She came after me with a wooden spoon, and I kept running. Sport got between us and snarled at her. She stopped in her tracks and yelled, "You're gonna have to come inside sooner or later!"

I didn't really think much about it, and in fact, forgot the whole thing. But after a while, I got thirsty and went inside for a drink. Mother hadn't forgotten, of course, and was patiently waiting for me. What a licking I got.

About two weeks later Sport mysteriously disappeared. I looked for him everywhere, calling his name and looking in all the favourite haunts that he and I had discovered. When I asked my dad what had happened to Sport, he said the wolves had got him, but I didn't believe it; I got the feeling he wasn't telling me everything. I never found out what happened, and I still resent him for that. And even now when I remember Sport, I feel the loss of my close friend. Funny how something like that can stick with you after all these years.

***

IT WAS EARLY SPRING. THE WEATHER WAS fairly mild, but there was still a layer of slushy snow on the ground. Marlene came home from school full of excitement because lots of the other kids were going with their parents to the Brandon Winter Fair. Her friend, Juney Taylor, had asked her mom and dad if Marlene and I could go with them. Mr. and Mrs. Taylor said that was okay with them, provided our parents agreed. When Marlene asked our parents if we could go, they said no. Their explanation was they could not afford for all of us to go, and they didn't want us to go with the Taylors. Our family had no money, and our parents were concerned about what the community thought of them, so they decided it was easier to just say no. Marlene was devastated. She cried and cried. By this time, I had gotten into the act, too, and we put on quite a performance. We were crying loudly and bemoaning our pitiful state.

And then a big surprise when my father finally decided that we *would* go. He started the old 1940 Chevy and drove up to the front of the house. Suddenly realizing she did not have her hairbrush, Marlene ran back into the house to get it. When Dad saw that, something snapped in him, and he flipped into one of his totally uncontrolled violent episodes. He revved the car and started spinning around in the yard, throwing mud and snow everywhere. Mother ran beside the car, grabbed onto the door handle, and screamed, "Stop, Bill! Stop!" Losing her footing in the slushy snow, she fell, fortunately sliding out of harm's way. He probably didn't know what she was doing, and if he did know, he probably didn't care. He roared off across the field and down the road, disappearing from sight. I don't know what Mother had in mind by doing that. Was she trying to stop him? Or was she going to take off with him and leave her five- and six-year-old behind? She certainly didn't seem concerned about us, neither at that moment nor later. What I remember best after all these years is her reaction. She stumbled around, not knowing what to do, before going inside without saying a word to us. Marlene and I watched everything from the steps at the front of the house.

I also remember the expression on Marlene's face, the look of horror, fear, and terrible guilt. I was stunned and could hardly move. My knees felt like water. When my big sister started to cry, I did my best to comfort her. Mother stayed in the house and we both felt so abandoned. That incident is a very clear memory, and I know Marlene blamed herself for what had happened. I also believe she carried that burden for the rest of her life.

Eventually, having got over his rage, Dad came back, and quietly said, "We'll go," but my sister and I refused to get in the car with him. Nothing more was spoken of the incident by any of us. And Marlene was never the same happy-go-lucky little girl again.

***

I REMEMBER MY MATERNAL GRANDMOTHER COMING TO live with us. She was a mail-order bride from England who had come to Canada and married my grandfather; she had a hard life. The family spent some time living in a granary on a farm in Saskatchewan during the Depression. Mother told us about all of them sleeping on the floor and having to step over the others to go outside to the bathroom. I only vaguely remember that grandfather, who died when I was about three or four.

I think Grandmother had an intellectual disability, but she also had spirit. Marlene and I were quite mean to her, as children can be, and we did things that I regret today. Eventually, she got angry with us and smacked us. Some details slip your mind, especially when you're not being very nice, but I do remember us running to Mother and complaining, and then Mother getting angry with Grandmother. Shortly after that, Grandmother moved to a little house in Kelwood, and died alone there about six months later.

There were five girls and two boys in my mother's family. Mother was the second youngest, and the older girls had left home as soon as they possibly could. I didn't know much at all about them as we had nothing to do with them, except for Mother's younger sister Dorothy. She and her husband Danny came to the farm one day and she was eating raspberries from a bush in front of the house. I remember running over there and telling her there were worms in those raspberries. The poor lady went behind the bush and threw up.

Years later, I heard the yarn that Grandfather had abused Grandmother terribly. One day, she had had enough, and when he fell asleep, she took her big heavy cast iron frying pan and whaled on him. When he came to the next morning, she told him if he ever hit her again, she would kill him. The story goes that he never did touch her again. I was very impressed with my grandmother after I heard that.

Apparently, Grandfather also had the habit of turning his children against each other. His two sons farmed across the road from each other, and when they were working in adjacent fields, they would often meet on the road and duke it out. They were both banned from the local hotel because if they met up there, they would start a fight. Not sure anyone ever knew what their arguments were about. Anyway, that was my mother's family. There are stories, too, about my great grandfather, but they are pretty bad, so I'll leave those for another day.

Mother told me that my grandfather wanted to give everything to her when he died, but my father talked him out of it. She was quite annoyed when my dad did that, but he knew her family, and I'm sure he was thinking of our safety. He didn't say anything about it to us, but I could sense him tense up when mother's family was around. I'll never forget the day two of her sisters drove onto our farmyard. They were just getting out of the car when Mother came running out of the house yelling, "Get the hell off my place!" I don't know why she did that, but I believe it had to do with my grandfather's estate. I remember the looks of astonishment and disbelief on their faces. That was my real introduction to my mother's side of the family. I got to know some of my cousins later in life and heard many more interesting stories.

Years later, Dorothy and I took Mother to her brother's funeral. She sat for a long time with one of those sisters and appeared to have a great time. On the way home, I suggested to her that she should go and visit that sister, but she wasn't about to do that. Then I suggested that she write her sister a letter instead, but she responded, "She isn't worth the price of a stamp!" And that was that.

When that happened, I should have seen what was coming. Eventually I had a similar experience with Mother when I didn't see what was coming, but it wouldn't have mattered anyway.

Bill at age seven.

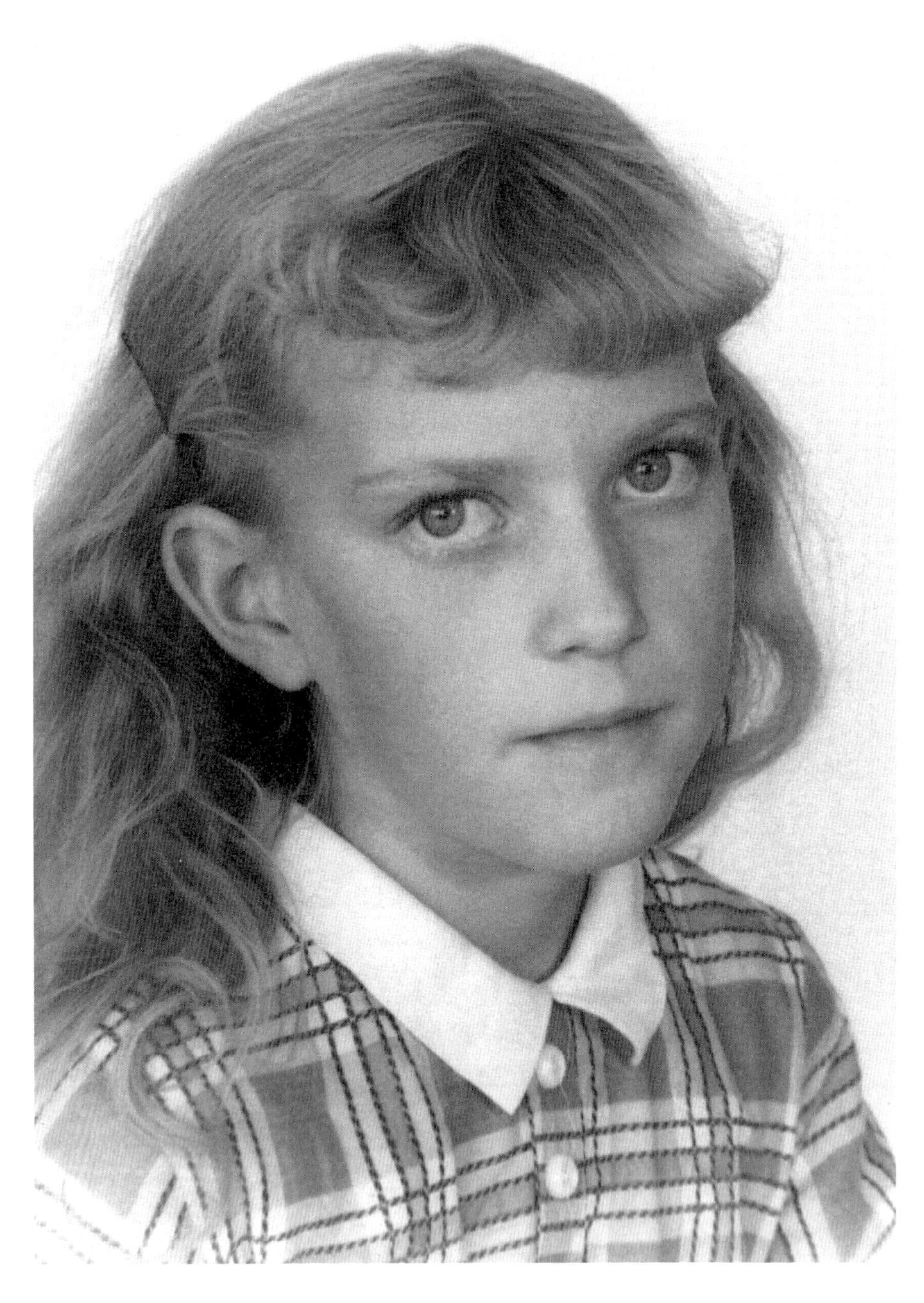

Marlene at age eight.

CHAPTER THREE

A Time of Learning

OUR NEXT MEETING TOOK PLACE ON JUNE 17, 2004, at the Big Island Community Hall. The first order of business was to name our group. Several ideas were bantered about, until someone came up with the idea to call ourselves the Concerned Citizens of Big Island—CCBI for short. Little did we know that our group would continue in this struggle for more than fifteen years. Even if we did, we would have done it anyway. For most of us there was no other choice but to enter into the conflict for the sake of our community, our children, and our grandchildren.

Our second order of business was to prepare a petition to stop expansion of the hog barn. It was signed by all of us at the meeting and circulated to others in our community.

People then reported on their research. Ducks Unlimited Canada informed us that phosphorus is of bigger concern than nitrogen in terms of spreading the manure on the land. They went on to say Manitoba has some of the toughest policies on manure management on paper, but does little to enforce them. The hog industry is having a hard time making a profit, so it is focusing on Hutterite Colonies due to their cheaper labour costs.

A professor from the Faculty of Agriculture and Food Sciences specializing in soil chemistry told us that phosphorus is added to the feed, which means that pig manure is very high in phosphorus. He said that nitrogen and phosphorus can saturate the soil down to a level of six to eight feet. Then he talked about manure storage facilities with clay liners, and indicated that nutrients can move through the clay into the surrounding soil and water. He said that clay can crack, and manure storage facilities can and will leak. He also mentioned a number of times that a hog operation should not be built over an aquifer where people depend on private wells for their drinking water. We found it all very disturbing.

A fact sheet prepared for the Center of Environmental Research at Cornell University in 1985 by McCasland, et al.,[1] says that nitrogen creates a potential health problem primarily for infants. It interferes with the blood's ability to carry oxygen, and can also create cancer-causing substances.

When I met with the provincial groundwater geologist from Manitoba Conservation, we looked at profiles of the bedrock and places where the bedrock came to the surface. We covered a lot of ground that day so I could show him where uncapped wells were located near the manure storage facility. We also talked about the direction of the flow of water underground, which is basically the same as the water on the surface. This was the first time dealing with a government official, and I have to say what I found was a solid adherence to the party line. I wasn't sure what to expect, but I soon found out. He was not going to commit to discussing concerns that might go against the government policies on hog barns. It seemed to me that there were clear and present dangers associated with this operation, but he would not have any of it. I felt frustrated and angry. It didn't matter what was said; it was pretty clear I would come up against a brick wall with these people.

I had many similar experiences with other government officials. We were told by former provincial conservation officers that their role was to assist producers to find their way around the government regulations, so it was not surprising that we got little sympathy. Even though another

1 Margaret McCasland, Nancy M. Trautmann, Keith S. Porter, and Robert J. Wagenet, "Nitrate: Health Effects in Drinking Water," *Cornell Cooperative Extension*, December 1985, http://psep.cce.cornell.edu/facts-slides-self/facts/nit-heef-grw85.aspx.

member reported that some local municipal officials were supportive of our position, I could already see that if we were going to protect our community, it was up to us and us alone. And I realized even more the importance of resisting this expansion. We met again on June 30 to finalize plans for the community meeting in Big Island on July 6.

***

IT MIGHT BE HELPFUL AT THIS POINT to explain that the way manure is spread depends on the type of farm operation.

Solid manure from straw-based operations is simply tossed out into the fields and then worked into the soil. The manure is generally spread over all the land a farmer has so as not to saturate the amount of nutrients in any particular area.

Liquid manure from factory farm operations is spread out from the manure storage facility—usually up to a 3-mile radius—which means there is a higher concentration of nutrients in a much smaller area. Hoses from the facility are connected to a cultivator that injects the manure into the soil in the fields, from near the surface to a depth of around 12 inches. When liquid manure is spread in areas where there is little or no topsoil, it is sprayed on the surface. And when manure is spread in wet conditions, there is the likelihood it will run into ditches, drainage systems, and watercourses.

The hog industry in Manitoba is predominantly factory farming. They store liquid manure in storage facilities and inject it into the soil once or twice a year, depending on the size of the operation.

In Manitoba, the permitted levels of phosphorous, nitrogen, and other nutrients are too high, which means plants are unable to deplete it. As a result, those nutrients either run off the surface into streams, rivers, and lakes, causing algal blooms (phosphorus), or make their way down to the aquifer and into the groundwater (nitrogen).

***

IT WAS ANOTHER CLEAR, BEAUTIFUL SUMMER EVENING when Dorothy and I headed over to the community hall in Big Island for the meeting. In our notification about the meeting, we had listed the speakers. I was to speak on behalf of CCBI about our concerns with the present hog operation and future expansion. The provincial manager of environmental livestock manure with Manitoba Conservation would talk about the manure management regulations. The director of the Interlake-Eastern region for Manitoba Agriculture would discuss the Technical Review process, and people from the Department of Agriculture would determine the suitability of this site for this hog facility. The provincial groundwater geologist from Conservation would talk about the aquifer, and the Chief Medical Officer from the Interlake Regional Health Authority would address health concerns around hog factory farming. Each speaker was given ten to fifteen minutes for their address, and questions would be answered later in an open house format where people could personally talk to each individual presenter. I had learned this technique as a principal when conducting public meetings at my school after some inmates escaped from the nearby federal penitentiary. As you can imagine, some of those meetings got quite intense.

Several days before the meeting, I had talked to a Portage area farmer who was a member of Hog Watch. He offered to come and speak at our meeting, but since I had already sent out a notification of the meeting, I asked him to speak at the end. The colony received the community newsletter about the meeting, as did everyone, and did not ask to be included on the speakers list.

There were several hundred people in the community hall that evening. Each government official got up and spoke. To listen to them, you would have thought that the hog industry was the best thing that ever happened to Manitoba. I wondered if they had ever considered what their jobs actually were—to protect the public and the environment of this province—but we were to find out otherwise. We were particularly disappointed in the representative from Public Health. Then our Hog Watch representative got up and took on each speaker individually. He clearly explained what the problems were and what we were likely to face down the road; unfortunately, he has been proven mostly right. As he continued talking, we could see each official's face getting redder and redder. The crowd was with him and began to react with some mirth at each point. I was really enjoying this.

One of the things we again raised at that meeting was the issue with manure from the colony's existing biotechs leaking into Colony Creek. Biotechs are shelters open to the air with straw-based manure management systems. They work well if operated properly, but the colony's ten biotechs were all built on sloping land, and the recent excessive rainfalls had mixed with the solid manure and leaked into the groundwater. That then drained directly into Colony Creek, which flows into the Assiniboine River. Once again, the colony said that would all be taken care of with the new development, but given their past record of 'taking care', we were not convinced.

When Peter, the colony business manager, realized the meeting wasn't going very well, he asked if he could get up speak after the last presentation was finished. I told him no, that he'd had his chance to be on the speakers' list and hadn't contacted me. Instead, I moved the meeting into the second part where people would have an opportunity to ask questions of each speaker.

The head of the Pork Council, who was there with Peter, came over and angrily and aggressively dressed me down for not letting him speak. Tom, a senior member of our group, came up to him and told him in a tone and volume that was very impressive, "You've had your meeting. This meeting is ours, and we'll run it the way we want to." And that was the end of it. Peter calmed the man down and sat at his table to talk to people who came to ask questions. I was ever so grateful to old Tom. This would not be the last time that I was attacked like that in a meeting, but as time went on and I learned more, I began to give as good as I got.

It was clear to all of our group that we had the support of the community to continue to oppose this proposal. It felt so good to get lots of encouragement from the people you are representing. I could see that I was going to enjoy this and began to look forward to continuing.

When CCBI met again on July 13, we agreed we needed to collect some more evidence to validate our concerns, and we sent a sample of water from Colony Creek for testing.

I met one of my neighbours, Old Joe, for the first time. He lived about a mile from me and less than a mile from the colony barn to the northwest. Joe was a retired commercial pilot and owned a small plane. I asked him if he would overfly the area where Colony Creek was being polluted so we could take some pictures. He was game, and Harold volunteered to go with him to take the pictures.

We had already reported the leakage of manure into Colony Creek to Conservation, but it wasn't until we took the aerial photographs that we discovered the extent of the situation. When Conservation visited the colony, they wrote an order to stop the leakage, and gave the colony until August 16 to correct the problem.

Conservation also discovered that the colony did not have a permit for the human waste lagoon that treated their sewage. However, rather than make them shut it down, the colony was simply told to get an engineer's report to bring the lagoon up to current standards. Of course, since they didn't have a permit, there was no reason for the colony to send in the required yearly reports on the lagoon, which meant no follow up from Conservation.

The colony had also not been filling in the Manure Management Plan (MMP) that was required by the government for an operation over 300 animal units (AUs). They were told to complete one and submit it twice a year. At that time, we could still obtain the producer's MMP from Conservation, and when we got a copy, it indicated there were 658 AUs in the hog operation. An AU is actually a measure of manure being produced, and one AU can represent twelve to twenty pigs, depending on their size. Their request for expansion had opened up a whole can of worms for the colony.

We realized we needed to talk to the councils of the three rural municipalities (RM) affected. It's all very confusing, so stay with me. The colony and the barn are located in the southeast corner of the RM of Muddy Woods. Within half a mile to the south is the RM of Greenland, and half a mile to the east is the RM of Snow Valley—the one in which Dorothy and I are located. Of the twelve homesteads within a mile of the proposed barn, six are in Greenland and Snow Valley. The Village of Big Island is just over a mile away and is also in Greenland and Snow Valley. Over the years, as I've gotten to know more about the situation in the province, it is apparently a common occurrence to locate a barn so that most of the people downwind are located in a different RM to the one where the barn is located. That way, most of the people affected have no political connection with the council. We knew we were going to have to talk to all three RMs about the situation.

On August 10, we met up with the Muddy Woods Council. We showed them our evidence, talked about our concerns, and handed over the petition we had started circulating in the community back in June. Sam, the hog barn manager from the colony, and Peter, the colony's business manager, were present at the meeting and were allowed to speak. The discussion became a debate until the reeve ended the meeting. It was apparent that the reeve did not know what he was doing or how to conduct a meeting, or perhaps that was his intention all along to prevent us from making our arguments. Nevertheless, this meant there was no opportunity for the council to ask questions. We certainly felt powerless and frustrated in that situation. They promised to consider our concerns if and when a formal application was filed with the RM of Muddy Woods. Shortly after that, we made a presentation to the other two RMs, both of which expressed their concerns to Muddy Woods.

Old Joe and I met with Peter and Sam at a local restaurant on August 24. After the meeting at Muddy Woods, we wanted to open the lines of communication. When we walked in, you could see they were loaded for bear. At first, they were defensive, but we assured them that we were representing Big Island and expressing the community's views as we saw them. We made it clear that the issue was the proposed hog barn expansion, not past disagreements or discrimination. Peter told us that the number of pigs in the final phase of this expansion would number 28,000. This represented more than 1,600 AUs—almost three times the amount of 658 AUs reported on their 2004 MMP.

Joe paid for the coffee, and the meeting seemed to end on a positive note, with them agreeing to meet with us again in the future. Despite the situation, I always enjoyed those meetings, along with the joking and bantering that went on, because it felt like the good old times before the expansion was on anyone's radar. Even at that early stage, however, we understood full well that the colony leaders were only telling us what we wanted to hear. I'm quite sure they were pleased with themselves and laughed at us all the way back to the colony. I believe they had a plan and followed it from the beginning. I'm sure they had coaching from somewhere.

We met again with the community on September 15 to report on our progress, and began to prepare ourselves for the results of the Technical Review and the possibility of a Conditional Use Hearing. The Conditional Use Hearing was an opportunity for the community to voice their concerns on the project. We asked the community for its support and involvement should that happen. We began gathering allies—the Vintage Locomotive Society whose old train brought thousands of people to Big Island every summer, the Prime Meridian Trail with its seventy-five-mile non-motorized trail that begins in Big Island, the RM of Headingley, City of Winnipeg West End councillors, and the Friends of Sturgeon Creek. A second local association was formed to oppose the barn—the Mothers of Young Children of Big Island.

I was not looking forward to a hearing. I could see we were outnumbered by the colony, feed companies, and government agencies, but we felt we were ready. We just never expected what happened next.

CHAPTER FOUR

Going to the Dogs

DAD'S OLD 1940 CHEVY CAR WAS A dark blue two-door business coupe. He would park it just outside the gate on the driveway that went through the bush on the farm. I was fascinated with that car. To start it you would push down on a button on the floor. You didn't need to turn on the key to make it work. I discovered that I could put it in first gear, push down on the button, and make it move.

Of course, I knew I shouldn't be doing that because I would wait until Dad was not around before I would go and play with the car. I remember him looking at the car afterwards with some confusion. He had parked it ten feet away from the fence, and now it was right up against it. Not only that, but the six-volt battery was dead, and he would have to get out the crank to start the car. This went on for several weeks, and I remember him telling Mother that he couldn't figure out what was going on. What *was* going on is that I would get in the car, put it in low gear, and push on the starter button. It would heave itself forward two or three times, which killed the battery in about five minutes.

The last time I did that, I was down on the floor pushing mightily on the starter button, and didn't notice my dad come out into the yard. Presumably when he saw the car move forward, with apparently no one inside, he ran over to the car and whipped the door open. And then he saw me. "Billy!" he yelled, although I think he was actually quite pleased that I'd figured this out. That story was told over and over with lots of laughter to the guys at the junkyard in Kelwood.

When I was about ten years old, Dad bought a four-door 1947 Chevy. He gave me the '40. Usually when I got a dime for my weekly allowance, I would spend it on a chocolate bar, a drink, or a comic book. But after I got that car, I could buy a gallon of gas with the ten cents. Can you believe that was the price of gas in those days? I put the gas in the old '40, and drove it all around the farm. Usually, it ran out of gas at the farthest corner, and the car would sit there until I could go back to town on Saturday and buy another gallon of gas. I even got it into high gear sometimes and bounced along at twenty miles per hour or more. I have always enjoyed speed.

***

IN THE SPRING OF 1955, WHEN I was six years old and going to Grade 1 in Glenallan School, Marlene and I came home one bright and sunny spring afternoon. A little ditch in front of our farm was full of water. We were both comfortable in and near water, and decided it was time to go for a swim. It didn't matter to us that there was still a snowbank stretching into the ditch. I took off my shoes and socks, rolled up my pant legs, and headed in. Marlene, a year older and much braver, stripped off all her clothes and dived straight in. I can still see that little bare bottom disappearing into the ice-cold water.

One day that same spring, I was coming home from school by myself when I decided to take a shortcut across the field to the house. I was well equipped for the journey, having just received a good pair of high rubber boots. But the mud was relentless. After it took one boot, I struggled along until the mud captured the other boot. Determined, I carried on to the point where I lost the first sock. By this time, Mother was yelling from the house that I should go back, but that didn't make a lot of sense to me because I was already two-thirds of the way across the field. However, having experienced Mother's wrath when I did not obey her, I decided I'd turn around and do as she instructed. Besides, this was actually becoming fun, and it actually became easier to navigate the mud when I lost the other sock. When I got back to the road, I considered my options. There was a snowbank in the bush that went around the edge of the field, and I thought if I ran fast enough the snow wouldn't seem too cold.

Despite obeying my mother, when I finally got to the house, she was angry anyway. After the inevitable spanking, I went inside and found another pair of socks. When Dad got home, he hitched up Trigger and Buttermilk to the high-wheeled wagon, drove them across the field through the mud, and rescued my boots *and* my socks. Another story for his friends.

The other machine I liked to play with on the farm was the old John Deere tractor. I discovered that if I put it in first gear, disengaged the clutch, and ran on the belt pulley, I could drive it around the yard. I also discovered I needed rubber soles on my footwear for this to work. Dad didn't seem to care because he couldn't afford to buy the gas for the tractor and didn't use it much anyway. This went on until I drove it into a mud puddle in the yard and got it stuck. It was out there until he wanted to use it again, when he simply waded into the water, started the tractor, and drove it out. After that, I moved on to other things.

I guess I have a mechanical bent like my father and grandfather before me. My son is that way too; it clearly runs in the family.

***

WE GOT ANOTHER DOG THAT REGULARLY FOLLOWED me to school and waited outside till school got out. We were walking home one day when suddenly it started writhing, yelping, and biting its stomach. It was awful! I don't have words to express how that affected me, but I remember the feelings of horror to this day. There had been a skunk hanging around the school, and someone had put out strychnine. I ran home to get help, but when we came back the dog was dead on the road. We gathered it up, took it home, and buried it.

I don't remember that dog's name. I think the trauma of that event wiped it from my memory. You don't forget a scene like that. Thinking about that experience now, I can't help but wonder what my dad saw and experienced during the war that I still don't know about. Probably things far worse and involving a lot of people. I can't even imagine what that was like, but I'm sure it changes people, and not for the better.

That winter, my grandfather, who had finally heard about my dog Sport, sent me another dog. He put it on the train in Ninette, and they forgot to take it off at Kelwood. The pup went on to Yorkton, Saskatchewan, but eventually, the railway realized their mistake and sent the dog back to Kelwood. That poor dog had been on the train for about ten days. Grandfather had written to tell me he was sending this Scotch Collie puppy, and I had been impatiently waiting for him to arrive. Finally, we got word from a neighbour that he was at the station. Dad started up the '47, and he and I headed into town.

The puppy was in a strong wooden box, and he wanted out of there in the worst way. The trainmen had been feeding and watering him, but he was very tired of being cooped up in that crate. I tried to get him out, but it was impossible. Dad told me to wait till we got home where he'd have the tools to get that crate open. The puppy and I got in the back seat of the '47, and on the way home, I discovered a crank on the floor in the car. I took the crank and reefed on that crate until I broke it enough for the puppy to get out. He was so happy to be out of there.

From that moment on, Shep was my dog. He became a great herding animal, and people from the district would come and get him and me when they needed to move a flock of sheep or herd of cattle. This made me feel great because people were relying on us to do a difficult job. I was learning the rewards of doing important and necessary work in my community, and receiving encouragement that I did not get at home. Shep seemed to know instinctively what I wanted. He was gentle with the animals, energetic, and clever. Sometimes, an old ewe would think she could get the better of him and lead him away, but he wasn't fooled. It would take a pretty smart animal to get the better of Shep.

Somehow or other we acquired a little female dog called Queenie. One day, Dad was mowing hay with the horse mower just in front of the house. There were lots of mice in the hay, and the two dogs were hunting them. Queenie got a little too close to the cutting bar and a hind leg was cut off at the knee before Dad could stop the horses. He carried her to the house, and I made a bed for her in a big cardboard box. She started licking the wound and then rested quietly. I guess it had been a clean cut, and she kept it from getting infected. I kept an eye on her and the wound, but it healed nicely without any human intervention. She survived the experience and went on to be a remarkable little three-legged dog that raised a number of litters of Shep's pups.

By this time, Shep's reputation in the district and elsewhere had grown far and wide. People would come for miles to get his puppies. My parents wouldn't charge for them, but the people who came would often give me five dollars because they knew that I had trained him. As you can imagine, I picked out the best ones for those people who offered me money. I had become an enterprising businessman. My work with the dogs was paying off.

CHAPTER FIVE

The Bit in Our Teeth

THE *STONEWALL ARGUS* HAD ATTENDED OUR PUBLIC meeting on September 15, 2004. They interviewed Peter and wrote an article in the September 17th edition that made clear his contempt for us and our position on the proposed expansion. He referred to us as dreamers, and while I know he didn't mean that as a compliment, I liked being thought of as a dreamer. I had been innovative in my job and had been able to bring lots of my dreams to fruition. Peter, too, is a dreamer and an innovator. Often if you meet him at the mailbox and ask how he is, he'll respond, "Just living the dream!" This happened when I ran into him just the other day.

Several of our members were quite indignant about his remarks and responded with an excellent letter to the editor. We knew that Conservation was going ahead with the Technical Review process and that the next step would be the Conditional Use Hearing. We were somewhat dismayed that our efforts to expose the colony's apparent disregard for the environment and the obvious concerns of the community seemed to make little difference to government officials. We were pretty naïve back then and somehow still wanted to believe the government agencies were there to look after us. They certainly led us to think that. Apparently, the *Winnipeg Free Press* kept an eye on the local papers, and when they called and wanted an interview, I asked for a chance to consult with Joe.

Joe lived a mile west of me on ten park-like acres. He had won an award from the rural municipality (RM) of Muddy Woods for his beautiful yard. Old Joe and his son Young Joe, who lived next door, were about a mile from the colony's manure storage facility. They would get the smell from the barn when the wind was from the southeast or from the manure storage facility when the wind was from the southwest. I drove over to where Joe was working in a garden.

"Hey, Joe, the *Free Press* wants to interview us about the hog barn. What do you think?"

Joe looked at me quizzically. "Why wouldn't we do it?"

"Because everybody who reads the *Free Press* will know that you live at Big Island and have a small aircraft." I was beginning to appreciate this man more and more. He had principles, and he had guts. When I was up against it, I knew he was in my corner. You know you have a true friend when that happens.

"Bring it on!" he said.

I went home, called the reporter, and told her we'd like to do the story. The next week the story appeared on the front page of the October 4th edition of the *Winnipeg Free Press*. On the second page was a picture of the manure spill that we had reported to Conservation back in the summer. The spill spread across the entire picture. Peter didn't have as much to say this time around. I'm not sure what effect that article had, but given the comments we heard afterwards, I do know that the government agencies and all three RMs we were dealing with paid some attention.

The colony hated appearing in the newspaper. There is plenty of rivalry among colonies, and the others often revel when one of their numbers is embarrassed in the paper. There is a lot of competition among them.

***

ON OCTOBER 13, WE HAD ANOTHER COMMUNITY meeting. The Mothers of Young Children of Big Island made a powerful and emotional presentation to Muddy Woods Council. I reported that after the article appeared in the *Winnipeg Free Press*, the Minister of Agriculture, Rosann Wowchuk had called me, but that conversation had been basically about damage control. We also talked about the articles that had appeared in the local papers. Jim Rondeau, the MLA representing the Headingley area, had called, as well as Wilfred Taillieu, Mayor of Headingley. We were in contact with the city councillors in St. James, Brookside, and St. Charles wards, all downstream of the hog barn. The article put us on the provincial stage.

There had been problems with smoke from the colony burning stubble fields that week, and when I asked if the community members wanted me to report on it, their response was an enthusiastic yes. We definitely had the bit in our teeth now. We talked about the colony's Manure Management Plan. We had information from the plan that they had been required to write back in July. Up till then, they had not been submitting those plans even though they exceeded the 300 animal unit (AU) requirement. We shared the Impact Analysis of intensive livestock operations on Manitoba rural residential property values that had been prepared by Royal LePage for Manitoba Pork. In the report, an Alberta tax appeal board ruled that properties located within a two-mile radius of a factory farm could see their values reduced by fifty percent.

We concluded by saying we had not heard anything about either the Technical Review or the Conditional Use Hearing.

After the meeting, I got in touch with Conservation to find out the current state of affairs regarding the human waste lagoon at the colony. The colony was allowed to discharge the sewage from the lagoon, but was then required by Conservation to have an engineer's assessment done on the lagoon. When the colony refused, Conservation informed me that they were initiating legal action to force the colony to comply. That whole process turned out to be quite the story.

Over the next few months, I checked several times with Conservation on the progress of the Technical Review, but each time was told it had not yet been released to the public. Around March 1, 2005, I went to the mailbox and found a notice from the colony stating that after extensive consideration and consultation with the RM of Muddy Woods and our community, the colony would not be increasing the number of AUs on their farm. It went on to say that the RM had issued the colony a permit for the construction of a new weanling and finisher facility that would replace existing out-of-date facilities.

Upon hearing that news, I allowed myself to be ecstatic, even though I had misgivings. You've heard the expression "waiting for the other shoe to drop"? Well, that feeling was there for sure, but I still put out a newsletter to the community thanking them for their support. One group that was involved with us, the Mothers of Young Children of Big Island, decided to disband. We all assumed that the battle was over.

I read the notice over the phone to Dorothy's brother, who simply said they were "good words." Because he was familiar with the Hutterite people's culture, that made me a little uneasy, but at the time, the significance of his words escaped me. Later, I knew clearly what he meant.

In their mail-out to the community, the colony mentioned that they had obtained a building permit for their new facilities from the RM of Muddy Woods. I asked the RM for a copy of that building permit, which of course is public information. It gave the colony permission to replace their existing hog barns with a building approximately 80,000 square feet in size. It was valued at $1.5 million and Muddy Woods charged a fee of fifty dollars for the permit. Later on, when we were having problems with truck traffic on the roads, we joked that fifty dollars would pay for about eight feet of dust control. It was at this point that my brother-in-law's comments about those "good words" were becoming clear. Well, at least we had four months of thinking that we were out of trouble.

Along with the building permit, we received copies of information from DGH Engineering, which described the current number of hogs on the farm as being 889 AUs—the number for compliance. That did not seem to compute with the Manure Management Plan we had obtained in the summer of '04 when the colony stated the number was 658 AUs. Keeping in mind that an AU represents approximately twelve to twenty pigs depending on their size, the difference between 658 and 889 could represent 3,500 animals.

***

ON MAY 24, 2005, WE HAD ANOTHER public meeting. I had received information from the federal government about the amount of space required to house pigs. According to my calculations, a barn of about 54,000 square feet would be sufficient to house the number of animals in the operation at the colony. Another issue discussed was the test wells around the manure storage facility. These wells are used to obtain samples of groundwater to see if there is any leakage of manure into the aquifer. The colony stated they were looking into test wells and odour control. No concrete promises on either subject were made.

I raised my concerns with the RM and the colony about the size of the barn. Sam agreed to meet with us at the RM office to explain why the barn was the size it was going to be. The economic development officer for the RM who had issued the building permit was late for the meeting. While we waited, Joe and I sat down with Sam to go over the plans. He described to us how much space was required for each aspect of the barn. He said they wanted a humane operation, so the barn was a little bigger than it needed to be. There was a room set aside for sorting out animals, one for sick animals, and another that would always stand empty for cleaning purposes. The explanation seemed reasonable at the time.

However, when the actual number of animals that were put in that barn was made clear, I wondered whether they gave up such things as office space, hallways, and so forth, or did they just cram as many pigs as possible into those rooms? I wondered, but I already knew the answer.

***

HERE'S A BIT OF INFORMATION ABOUT PIGS on some farms. In the literature I got from the federal government, the requirement is 11.3 square feet per adult pig. That's a space less than six feet by two feet. An adult pig takes up that much room when it lies down. Can you imagine living in an area that big for your entire life? It's called factory farming and it's obscene. How can we as a civilized society treat animals that way? The answer to that question is it's all in the interests of making money.

I once helped Dorothy's sister's partner haul pigs down to the U.S. When those pigs were loaded into the truck from a factory barn, they were squealing, biting at each other, fighting, and covered with scratches and scars. We hauled them down to a South Dakota packing house, and were one of ten semis unloading at the same time. The stream of pigs going to slaughter was like a river. This was in August, and a Bobcat was being used to haul any dead animals from the trucks to a dumpster. There were often two or three dead pigs in each truck. Our load didn't have any because my brother-in-law was careful to frequently give the animals water in that hot weather. The sights and smells were horrific.

I know pigs. I raise four or five every summer as food for my family. They have a room in the barn ten feet by ten feet, and they can go outside into a dirt pen that is sixteen feet square. They are calm, intelligent, alert, and happy, just like the pigs we raised at home on the farm. You can always tell a happy pig by the tightness of the curl on its tail. I have some remarkable pigs here this year.

One evening, I heard a commotion in the pen and went over to see what was going on. Three of them were standing nose to nose, facing each other in a circle. As I watched, they started shuffling around, squealing and grunting with pleasure as they kept their noses pointed at each other. It reminded me of square dancing. I have never seen that before. I often hear them playing, and wonder if they sense I'm trying to do something for their kind.

One year, I did buy some small pigs out of a factory barn. The farmer did not charge me for one of them because it was shaking so badly it could hardly stand up, and he was going to destroy it anyway. I took it and put it in the pen outside with the others. After about two weeks, it had stopped shaking and seemed fine. I think the only thing wrong with the animal was the stress it had endured in that barn. It was heartbreaking to watch it suffering. Even after it recovered, I felt such anger about the conditions in those factory farms. Talk about injustice. Don't get me started.

Bill and several of his five pigs.

CHAPTER SIX

The Best Grandfather Ever

DAD HAD FAMILY IN SOUTHERN MANITOBA. THERE was Evelyn, his older sister, who had four children, some older and some younger than Marlene and me. There was a younger sister who had several children younger than us. And then there was my grandfather, Will, who lived in Ninette, Manitoba. After his wife, my grandmother, died in 1952, Evelyn became the matriarch of their family. This was nothing new for her because at age 16, she had already taken on much of the responsibility of looking after the family when her mother suffered a heart attack. Evelyn was this wholesome, practical, "salt of the earth" kind of woman. I'm sure you know the type.

My grandfather lived in a small house in Ninette. There was something about him that fascinated me. When Aunt Evelyn began inviting Marlene and me down for several weeks in the summer, I insisted that I wanted to spend a week of that time with Grandfather. He had emigrated from England to Canada in 1903, worked for Evelyn's grandfather, and married her mother. I loved Grandfather's smell and remember it today—old wood and tobacco. I find it very comforting. Grandmother had been a teacher, and when her mother had suffered a heart attack, she came home to look after the family and the hired man. From all accounts, she was a wonderful woman and much loved by her family.

Grandfather homesteaded at Lonely Lake, Manitoba, in 1918, lived in a cabin with his wife and four children at the time, and proved up and got title to the land in 1923. That in itself is quite the story, but I'll save it for another day. While they were there, Grandmother inherited some land near Killarney. They bought a quarter section of land adjoining hers and moved back to southern Manitoba after selling the farm at Lonely Lake.

Because Grandfather had training as a steam engineer in the British Navy and was mechanically adept, he built a seed-cleaning plant and cleaned seed for the farmers in his area. He farmed his own half section with mules. He liked mules because they were smarter than horses and more determined, or, shall we say, stubborn. That was probably why he liked *me* so much. This was where my father grew up and went to school in the Hullett district.

By 1933, in the midst of the depression, they were unable to hang on to the farm at Killarney. Even the land that Grandmother had inherited was heavily mortgaged. Grandfather traded a team of horses for a quarter section in the hills northeast of Ninette. He and my uncle built a log cabin there in the winter of 1933/34, and in January 1934, when Dad was fifteen, the family moved. They lived there till 1949—the year I was born—when they sold the farm and bought the small house in Ninette. Dad became friends with some of the people that lived in those hills—remittance men who taught him to hunt deer. When I was sixteen, he taught me to hunt deer; he was an excellent hunter and taught me well. From time to time over the years, we went back to the farm at Ninette where the new owners allowed us to hunt. That was where I shot my first deer.

When I was eight years old, in 1957, I insisted that I stay a week with Grandfather. My aunt couldn't figure this out. Why would I want to stay with that old man when I could spend the time with her children who were about my age? He wasn't all that excited about it either, but I persisted. I was very happy when he finally agreed, although for the first few days he was a bit sharp with me.

Grandfather had converted his living room to a wood shop, and there were some beautiful thin pieces of wood lying about under the saw. I asked him for some of them and built myself a glider. I was playing with it in the backyard when suddenly a gust of wind took it over the fence into the neighbour's corn patch. I got scared and wasn't sure what to do, and being eight years old, I did nothing. I forgot all about the glider until several days later when I saw the neighbour and Grandfather talking over the fence. The neighbour had my glider in his hands. My heart sank into my boots as Grandfather beckoned me over.

"Is this yours?" he asked.

I said yes because I knew he'd watched me build it.

"Why didn't you tell me about it?"

"I didn't want to get into trouble with you, Grandfather." I spoke in an eight-year-old boy's trembling voice. I believe that was one of the few times I ever saw my grandfather speechless.

The neighbour said something like, "You can play in my yard anytime you like, Billy. You're a good boy."

Things were easy between Grandfather and me after that. He knew then I was committed to a relationship with him. It turned out to be one of the most important relationships in my life.

***

GRANDFATHER AND I SOMETIMES WENT DOWN TO the beach at Pelican Lake in Ninette. He would tell stories and we would fish, and sometimes beachcomb with another gentleman. That fellow was about Grandfather's age, and had a little Ford 9N tractor that pulled a wagon into which we tossed weeds and other debris we had picked up off the beach. I thought that little tractor was just the best thing I'd ever seen. I vaguely remember sitting on my grandfather's knee, steering the tractor while he drove.

Grandfather had adapted a rowboat and put a sickle bar mower attachment on it. It was pushed along by a little outboard motor, and he would cut the reeds along the beach. One windy morning when he went down to the beach to make sure everything was okay, he found a sailboat had slipped its moorings and was pounding on the rocks. My grandfather—seventy-five at the time, ever the seaman—waded into the water, pushed the sailboat out, and moored it to the dock. The owner was so grateful that he took Grandfather and me sailing that afternoon on Pelican Lake. I'll never forget the beauty of that afternoon and my admiration for my grandfather.

Once or twice Grandfather took me to downtown Ninette and bought me liquorice ice cream. I still remember how good it tasted. My aunt was worried about me and came over to see if Grandfather was feeding me properly. She glowered at him in his kitchen and asked him what we were eating. I stared back at her and said we ate eggs and corned beef. She was a little taken aback by that and didn't bother us anymore. In the evenings, Grandfather would tell stories about his time in the British navy and taught me cribbage. I played cribbage with him for quite a few years. At first, I seldom won, but towards the end I seldom lost. He liked that. I still have many great memories of the time I spent with him, and I realize now how important he was in my life. I tell his story, which I call *Tales of the Homestead*, on the Prairie Dog train here in Manitoba every summer.

Over the years, I've gone out to his place in Ninette a number of times, including about 15 years ago on a misty cool March morning. When I stuck my head in the open window, I immediately smelled old wood and tobacco—my grandfather's smell. I have experienced that smell several times since, and I believe my grandfather is with me when I sense it.

***

DAD HAD BROKEN A PISTON IN THE '47 and spent days honing out the cylinder to put in one that was slightly bigger. The car ran, but the problem was that if he drove over thirty-five miles an hour, it would start to knock.

We often went to Aunt Evelyn's in Margaret, Manitoba, for Christmas dinner, and in that old car, the trip took about five hours, one way. We'd start out at around eight o'clock in the morning and arrive there at about one o'clock in the afternoon. The doors on the '47 were not well sealed, and if it was blowing and snowing, the snow would drift onto the floor of the car in the back seat. It was hardly warm enough to keep the windshield clear. My sister and I curled up with blankets in the back seat.

When we arrived, the house was full of the beautiful smells of roast turkey and all the fixings. Everyone was there, including my younger aunt, her husband and their children, Evelyn and her family, and my grandfather. We would sit down to a wonderful Christmas dinner, and afterwards, there were gifts under the tree for everyone. My father, who was normally silent, sullen and angry, was a different person—happy, jovial, with lots to say. I would look at him and wonder who is this man? He was never like that at home. Presents would be opened, a few Christmas carols sung, and then a snack for the road. We had a five-hour trip ahead of us through the dark, often stormy, cold night. We wrapped up in a blanket, keeping our feet off the floor, out of the snow, remembering the sounds, the smells, and good feelings of that Christmas celebration.

By the time we got home, the wood stove was burned out, and the house was cold and dark. I don't remember any conversation on the way home. It was like we'd gone from one universe to another. It was such a contrast. I generally don't care much for Christmas because I remember those childhood experiences arriving home to a stark, cold house.

***

I HAD THE SAME TEACHER IN GRADES 4, 5, and 6. She was an older woman, and I didn't like her much. I thought she had favourites, and I wasn't one of them. I remember one of her favourites, Gerry Taylor, bullying me whenever he could. It went on for a while, and when I complained to the teacher, it did me no good; in fact, it just got worse. Finally, I'd had enough. He was pushing me around, so I wound up my good left hand and punched him right on the nose. He started flailing at me, but I had lots of practice in protecting myself. He didn't carry that on for too long and went away to nurse his bleeding nose. I didn't have trouble with Gerry at all after that.

My dad was working for the neighbours to try and make a little money to buy essentials, like sugar and salt. I was nine years old, and he was expecting me to do my share of the work around the farm. For the most part, I was willing, especially when it meant driving the tractor or working with the horses. There were some jobs that he didn't want to do and expected me to do them.

One day, I told him I didn't really feel like pulling nails all day. Disdainfully, he told me to go get my sister because she'd do the job. He was quite demeaning about it. I called his bluff, and I went to get my sister. I told her, "Dad wants you in the old house and has a job for you." She looked at me, but she went anyway to see what he wanted. He didn't really mean it; he only intended to humiliate me. When that didn't work, he sent her away. He hardly talked to me at all, except for jabs he sent my way.

One evening, he told me to water the horses. The mare had developed a habit of jerking loose when you took both horses together, so he told me to take them one at a time. I was busy and had things to do, and I thought I could handle her. Besides, the mare got anxious if you left her behind. So I took them both out of the barn together, and sure enough, she jerked back at a time I wasn't expecting and got loose. It didn't really matter because she was tied to that gelding by an invisible string and wouldn't go anywhere without him anyway.

Dad was in the yard and saw this happen. It's hard to describe the sound that he made. It was something like a roar, a scream, a snarled curse, all blended together, and then he ran full tilt towards me. I dropped the lead on the gelding and headed at top speed for the closest bush. I got there ahead of him and crawled under a thick tree. It was quite dark, but he continued to hunt for me. I tried to control my trembling and kept very still. But when he turned to leave, I moved just a little bit. It was a trick, and he grabbed me and lifted me off my feet. I didn't feel anything. I imagine that's how a rabbit feels when it's caught by a coyote; the shock creates a feeling of numbness. I saw the blow coming, but I didn't feel that either. I did, however, feel myself falling into the snow, and then the world went black.

My father left me bleeding there into the snow. I'm not sure how long I was out, but when I came to, the bleeding had stopped and the snow felt good on my bruised face. I picked myself up, went inside, and crawled into my bed. Nobody paid any attention or spoke to me. My hatred for my dad was pure that night. Where he was concerned, things changed for me forever.

CHAPTER SEVEN

An Unbearable Stench

UNDER THE GUISE OF REPLACING EXISTING FACILITIES, the colony began preparing the site for construction. 2005 was a wet year, and the weather caused delays. Because it was so wet, I didn't even put a crop in that year—the first and only time that has ever happened. We knew we were living on borrowed time and I had a feeling of impending trouble. I believe that ever since the time my dad first assaulted me, I've had a sixth sense about troubles lying ahead.

About a mile southwest of the manure storage facility was the Grant's Lake Wildlife Management Area, a nature preserve that was an important stopover for ducks and geese every spring and fall. As many as twenty thousand birds migrated through each year. When they spread their liquid manure, the colony uses a big cultivator to inject it into the soil in the fields around that preserve every spring and fall. When injecting manure, the colony is supposed to have a setback distance from the edge of the fields to prevent the manure from going where it shouldn't. So it was quite shocking to see puddles of manure on the edges of the fields and in the ditches draining into the preserve throughout the fall of 2005.

Harold, his wife, his sister, and her husband were quite concerned about the situation. Many of the local people, including me, have hunted ducks and geese along the edges of the preserve for years, and valued it. Birds would nest there and raise their young, and there were many species of birds and animals in the area. Harold's sister took pictures and I reported the spill to Conservation.

The Conservation officer came out to investigate the situation and then went over to talk to the colony. That led to the colony being sent a warning letter about the incident. I asked to be kept informed of the consequences, and was stunned to learn that no charges were laid, and no fine was levied. I told Harold and his sister about the inadequate response, and they were equally disgusted. To my knowledge this was the first of a number of incidents and similar responses by Conservation. It is extremely discouraging when people who are supposed to be protecting the environment don't do their jobs.

***

THROUGHOUT 2006, CONSTRUCTION CONTINUED ON THE BARN. It was finally completed in October of that year, and on October 24, the colony held an open house to show it off. There were a number of people there when Dorothy and I walked in. Sam was waiting for me and gave me a personal and detailed tour of the facility. I asked him about holding pens and rooms for sick animals, but he didn't really answer those questions. He's pretty smooth, I have to say. The building was obviously state of the art, with the latest technology installed. It was totally automated with individual rations of feed for each animal determined by their weight. There were people from several other colonies present, as well as municipal officials and people like me from the community.

The Hutterites made a big show of the fact that this was the last day the public would be allowed in the building. They said this was to prevent the transmission of diseases that both humans and hogs share—diseases that can easily be transmitted between animals and humans. We believed the actual reason was that it facilitated their hiding the true number of animals in the barn.

Because there is a high risk of diseases getting into a barn and running rampant between animals in close confinement, there's an extensive use of antibiotics in the feed. One of the major concerns is the concentration of these antibiotics in the environment, and also that strains of bacteria are becoming resistant to antibiotics because of the addition to the feed.

In research presented at the London Microbiome Meeting in 2018, Nicola Evans, doctoral researcher in structural biology at King's College London, shared some of her insights into this. She said the most important thing to consider is that any time antibiotics are used, whether in animals or humans, you run the risk of selecting for drug-resistant bacteria instead of ones that can be controlled. Antibiotics should be safeguarded for use in both animals and humans to ensure they can be used for the treatment of infection in the future. Some researchers predict that by the year 2050, antibiotic resistance will lead to 10 million deaths every year, surpassing cancer as the leading cause of mortality worldwide.[2] Pretty serious stuff!

***

DOROTHY AND I KNEW THAT WE'D HAD a reprieve for the past several years while the barn was being built, and by Christmas time, we realized what we were up against. One very cold day while putting up Christmas lights on the outside of the house, the light southwest wind carried the smell from the barn. It was extreme, to say the least, and felt acidic in your nose and throat. I felt sick that all our hard work to prevent this seemed to have been circumvented by the colony and those experts and officials who allowed this to happen. We were driven indoors and had to wait until the wind shifted the next day before we could finish putting up the lights. There was a sense of foreboding that something was coming, but we didn't know how bad it would be until it happened.

We weren't the only ones experiencing the smell. My neighbour took some photographs that showed the steam from the barn coming out

2 Ana Sandoiu, "Drug resistance: Does antibiotic use in animals affect human health?" *Medical News Today*, November 9, 2018, https://www.medicalnewstoday.com/articles/323639.

of the vents on the roof, dropping to the surface of the ground, and drifting over into his yard. Those photographs told it all.

I made contact with the chair of the Farm Practices Protection Board (FPPB) and asked for direction on the process of making a complaint. The FPPB had been set up as an alternative to suing the producers. Along with the legislation the province had enacted, what we came to discover was that its real purpose was to make it almost impossible for anyone to take any action against the producer or the rural municipality. Nevertheless, after telling me the smell from the barn was not in their mandate, the director gave me information on how to go about launching a complaint. A number of us began gathering data on the smell.

Coincidentally, around that time, the province of Manitoba placed a moratorium on any further construction of barns in the watershed for Lake Winnipeg. That was in response to warnings from scientists about high levels of phosphorous in the lake and the resulting threat to the fisheries industry. Too little too late.

The Clean Environment Commission was then established and conducted hearings around the province. We presented at the hearings in Stonewall, Manitoba. Young Joe and I outlined the concerns that we had with this project so far. We also made eleven recommendations to the Commission and provided them with photographs and other evidence to back up our concerns.

Peter presented as well. He quoted scripture from the Bible implying those passages supported their farming practices. He talked about colonization and the exploitation of the land as if it was something honourable. He complained about people who lived on acreages in rural areas and their concerns about smell that interfered with their operations. He also talked about caveats where people agree not to complain about the smell, even though such caveats do not exist. That's because you can't sign away your basic rights, or the rights of people who may own that property after you. I felt sorry for him because it seemed he was out of his depth, and I also felt embarrassed because I knew he was a much smarter man than he appeared to be. I realized later he would often use this strategy to come across as a simple religious farmer when he was, in fact, a very clever and ambitious man. You have to be careful when you are dealing with those types. Don't be taken in.

Most of the environmentalists who opposed the hog industry in Manitoba presented at hearings in Winnipeg. We were the first anti-industry rural presentation. Various publications attended our event and wrote articles that were quite uncomplimentary to us. In response, I wrote a scathing letter to the editor of the *Western Producer* in which I tried to expose their biases. I didn't think they would publish it, but they did, and I got calls of support from across the province. That was nice!

The crux of our letter was that we were offended by the tone of their article, which suggested that our non-Hutterite group was only complaining about the colony's operation because we were anti-Hutterite. I hoped my letter would set them straight because nothing could have been further from the truth. Besides, comments like that only serve to distract from the real issue. As far as I knew, it wasn't Sam who said anything along those lines to the reporter; he had received his share of backlash from previous negative comments, and now seemed to be doing a better job with the media.

Our recommendations to the Commission were intended to address issues across the province similar to the ones that we faced, but did the province pay any attention? I think not.

***

ON APRIL 29, 2007, YOUNG JOE AND I met with Sam, Peter, and his brother Walter in a restaurant in the local village. The first thing we talked about was the smell that had been consistently bad and affected us every time the wind blew from the south-west—twenty times since December. Young Joe asked if the manure storage facility was covered; it had been stinking more that spring.

They admitted they had made some unengineered changes to raise the sides of the facility, and claimed it was part of a project to capture methane gases and get carbon credits. We believed those changes were actually to increase the capacity of the facility because of the number of pigs they had. Whatever the reason, because they hadn't notified Conservation nor had an engineering assessment done before making those changes, Conservation wasn't letting them put the cover back on.

I believed the heavy cover might have also contributed to a collapse of the facility's walls, and that was likely the reason for, or at least added to, the stink. When I complained to Conservation about the facility and asked about the cover, they confirmed that it would not be allowed back on until the colony had arranged for an engineering assessment on the construction.

Sam also told us they were phasing in a feed additive that would help to reduce odour. Later, when I looked for research on this, I learned that the additives actually had little impact on the odour; they simply caused the pigs to digest their food more efficiently. I suggested that the colony host a barbecue or open house for the community in the fall to share more information about the barn. There was no interest in our suggestion, nor did they want to create a newsletter together that could be distributed in the community. I even offered to search out money for research projects to reduce the odour, but they weren't interested in that either. They did, however, assure us that they were committed to addressing odour problems—something they claimed to have been working on since they started planning the facility. We eventually discovered why they were putting up so many roadblocks.

They also said they would be careful to only clean or vent the barn at times least likely to affect their neighbours; the smell was even more rank when that was being done. I do think there was some attempt to do this—specifically to clean or vent when the wind was blowing in my direction.

Another problem was they had a manure pipeline from the barn to the manure storage facility that would occasionally become blocked and cause a rupture, resulting in the spillage of manure. They told us they had contacted Conservation about it, that the problem had been addressed, and it should not happen again.

It was clear the colony was not happy with our presentation to the Clean Environment Commission. Our position had been that we wanted to talk about general concerns and recommendations for the province using our experiences at Big Island as an example, but the colony interpreted it as us picking on them. Clearly, they couldn't or wouldn't see either the bigger picture or the rancor that had developed between them and the community. They just considered us to be mean people.

Nevertheless, there was a general agreement to keep the lines of communication open, and Sam said he would attend our Concerned Citizens of Big Island (CCBI) meetings. Then we told the colony we were considering a complaint to the FPPB if there was no immediate resolution to the odour problem.

***

CCBI MET AGAIN ON AUGUST 15, 2007, with Sam and a representative of the feed company. We had talked to Conservation about the situation with the manure storage facility, and the director had assured Old Joe that the cover would be in place within ten days. That never happened and the stench was atrocious. As suspected, the wall of the facility had been damaged, but was now repaired. We didn't know then that it would be a long time before that cover was installed.

Sam and the representative of the feed company reiterated that they were experimenting with feed additives to reduce the odour. I tried to pin Sam down as to when we might see some differences in smell from these additives. Instead of replying, he snapped at me, so I turned to the representative from the feed company and asked him. Sam had a smirk on his face, and in the end I did not get a response from either of them. I found their behaviour to be quite arrogant.

After the meeting, Old Joe and I and another neighbour walked out of the hall together. Joe was frustrated and angry and thought that Sam had got the better of me in the meeting. He expressed his frustrations and said that we hadn't accomplished anything in our efforts so far and we should probably give up. I think he was really angry at the way

Sam had treated me, and while I appreciated his concern, I got a little annoyed with him.

"You don't know where we would be if we had not attempted to stop this development, Joe. We could be looking at a barn of more than 1,600 animal units that was being permitted to function. Can you imagine what the smell from a facility that size would be like? What you're saying is just not fair." The other neighbour who was present nodded her agreement.

I think Joe realized I was right. He never talked that way again and was solid as a rock after that. We were on our way to becoming fast friends who had a lot of respect for each other. It seemed we were cut from the same cloth. Over the next few years, Old Joe and I restored a couple of trucks and a tractor together. We flew together in his small plane, and he became a great friend.

As it turned out, that was the last meeting we were to have with the colony for the next twelve years. For me, it was the day the battle really got started.

CHAPTER EIGHT

Those School Days

THE WINTER OF 1956 AND 1957, MARLENE and I caught a ride to school with Mr. Moylan and his children. The Moylans were even poorer than we were, if that was possible. I recall how the children were teased at school. Thinking about it now, it was probably because they were Métis. They didn't attend all that often, but when they did, Mr. Moylan drove one horse in front of a homemade toboggan. It would have been about twelve feet long and three feet wide, perfect for snaking through the bush. On really cold days, we would climb in and go to school with the Moylan children, all of us warm and toasty under the furs that Mr. Moylan had loaded onto the toboggan. As the steam came off the horse's nostrils and its nose covered with frost, Mr. Moylan stood at the front, a silent figure, tall and hunched against the cold. The toboggan slid easily, whispering quietly over the snow. It seemed to me that those children went to school on the really cold days, and on warmer days, they were out trapping furs with their father.

Going to school in the winter was certainly an adventure. Another family, the Tengans, had a little van with a stove in it that burned coal. It was quite cozy in there and it was fun feeding the fire. In fact, we often got it too warm and had to open the back door to cool it off. They would come from the other direction to pick up Marlene and me and take us to school. Their van was pulled by a team of horses, and they had quite the trail across the field to our place. Less fun was when Dad would sometimes hitch up the team and sleigh to take us to school. The sleigh had an open grain box on it that wasn't nearly as warm a ride.

One day, it had rained and the snow was covered with a thin layer of ice. Dad was not there to pick us up after school so we started walking home. At one point, we saw him with the team and sleigh galloping along in the neighbour's field. The horses were breaking through the ice and doing fine, but just ahead of the horses was a gray wolf running on the surface. It couldn't get any traction because of the ice and couldn't get away because it was slipping and sliding. As I watched, I felt some sympathy for the animal as it tried desperately to get away. Dad chased it up to the edge of the bush and stopped the team about a hundred yards away. Being a good hunter, he knew the wolf would stop, turn, and look at him before entering the bush. He was ready and brought it down with a .22 rifle he always carried. We were about halfway home when he caught up with us, and there was the gray wolf in the grain box. He got a bounty for the ears at the municipal office and sold the hide to S.I.R., a fur company in Winnipeg. That wolf kept us in groceries for a couple of weeks.

***

WHEN THE RURAL MUNICIPALITY OF GLENELLA BUILT up the road to the farm, they made a great ditch along the road with a dragline; the ditch was deep and full of water. Because it was new, there were no weeds growing there, and it was great for swimming. In fact, Marlene and I learned to swim there. We spent many happy hours in the ditch that summer.

Father was a strange man sometimes, and would often get ideas in his head that were not very practical. When he decided I needed a boat, he immediately began building it for me. Mother thought that instead he should be cleaning the barn or working the summer fallow or making some money at the neighbours, but once he got started, there was nothing that could deter him from following it through to completion.

He cut down a big tree with an axe, chopped out an eight-foot section, and started hollowing out the interior to make a boat. It's pretty safe to say that I was probably the only prairie boy to have a dugout canoe.

When he had finished, we put it in the water. With a round bottom, it had no stability, and I couldn't keep it upright. We solved that problem by nailing a couple of one-by-fours across the canoe and putting a couple of two-by-four outriggers on each side. After that it worked great! I even fastened a sail to it at one point. Because it rode low in the water, especially with me in it, if the wind was the right direction, it sailed beautifully. When Andrew Cole came over from next door to play with me, we often sailed the canoe. My grandfather was quite impressed when he heard about it. Between Grandfather's stories and my experience sailing my dugout canoe, I've always had a love of boats and water.

***

THE OLD HOUSE WE LIVED IN ON the farm was drafty and not insulated. When Marlene and I were little, we'd stay in bed as long as we could until we thought Dad would be up and making the fire. When we went downstairs, we would have to break the ice on the water pail to get a drink.

Another memory I have about the house was something that happened the summer we first moved there. Mother, Marlene, and I were sitting in the living room. There was a large picture window facing south, and a wild thunderstorm was going on outside. Suddenly, a crack shook the whole house and a bolt of lightning struck a fence post on the other side of the yard. We were all very startled; we jumped and cried out and were quite frightened. The fence post caught fire and burned steadily all through the downpour that followed. That is the closest experience with lightning I've ever had. I have always had great respect for thunderstorms and I enjoy their unbridled energy.

In 1956, the local school board decided to build a new school. The old one had been built in 1896 and was getting pretty hard to maintain. It had a big wood stove in the centre of the room, and one of the parents would come early on a winter's day to light the fire. That stove was really handy for toasting sandwiches at lunch.

One of the bigger boys decided to play a prank. He drilled a small hole in a piece of firewood and put a .22 shell inside. He then stuck the piece of wood somewhere in the middle of the wood pile. I can imagine that boy wondering every day he came to school if this would be the day. As it turned out, the day the piece of wood was innocently tossed into the fire, he was home helping his dad on the farm. When the shell went off, it was like a little bomb in that stove. The lids blew off, and the school was filled with smoke and ashes amid the screams and shouts of the students and the teacher. We got to go home early that day.

Father bought the old school. He decided he would move it to our farm in the wintertime on sleighs that he borrowed from our neighbours. It was very low to the ground, and he realized he could not get under it to jack it up. I was about eight years old at the time, and had no trouble crawling underneath with a little hydraulic jack. I would raise it a couple of inches and block it up. Then I would crawl to another spot and do the same thing. As I crawled around in the dirt under the school, I found a pistol. I think it was about a .32 calibre revolver that was rusted solid. All the wood was rotted off the handle. My friends and I played with that revolver for years. I often wondered about the story behind that revolver, and at a school reunion a few years ago, I asked the others, but no one knew anything about it. I think that someone had to get rid of the gun and threw it under the school. My grandchildren don't believe me when I tell them I played with a revolver when I was their age.

Eventually, I got the building high enough that Dad could crawl underneath and take over. For the next month or so, he got it so he could put four sets of sleighs under it. By this time, it had snowed, and moving day arrived. The Taylors came over with their two big John Deere tractors. Four tractors, as well as our old G, hooked on. All was going reasonably well until the sleighs encountered a small blow dirt bank at the edge of the schoolyard. As the sleighs passed over this ridge, a loud cracking announced that several of them had been damaged. The procession stopped at that point, and the attempt was abandoned. We had some of the tallest poplar trees in the country in our bush, so Dad cut several down and fashioned skids to replace the damaged sleighs under the school.

Towards the end of winter, he was ready to try again. This time Dad hired a Cat, and the Taylors came with their two big John Deeres and hooked up one on each skid at the front of the building. With the Cat pushing behind, they moved that school to our yard. When they got there, the tractors stopped before the Cat did, and the forward motion caused the beams on the skids under the school to turn flat on their sides. With no support for the floor, the walls of the building pushed down, causing quite a buckle in the floor. More damage occurred when several pieces of siding broke where the Cat had been pushing against the building while it was being moved. And that is how it stayed, skids and all.

We lived in that school building for the next five years. The old house was abandoned and we used some of the lumber in it for other buildings. The school was certainly warmer than the old house. It had a row of windows facing south, so we had solar heat on sunny afternoons in winter.

I guess Dad was a bit slow in getting those sleighs fixed and back to the neighbours. When one of them, Mr. Cole, needed his sleigh, he got a bit impatient with Dad. Now, my parents were always careful to make sure the neighbours did not know what went on in our home, but that day, Dad got so angry, he could not hide it, and Mr. Cole got an accurate impression of my dad. After that, Mother became good friends with Mr. Cole, probably to spite Father. Mr. Cole was our closest neighbour, and while I'm pretty sure he also heard and saw things that went on, he was not a gossip and never spoke of it until years later to my friend, his son, Andrew. The best secret is the one revealed when the time is right.

CHAPTER NINE

The Truth Revealed

WE THOUGHT WE SHOULD MEET WITH THE rural municipality (RM) of Muddy Woods again to inform them of the latest developments and our plans to go to the Farm Practices Protection Board (FPBB). I wrote them a letter on August 21, 2007, requesting that we appear as a delegation to meet with them, but received no response. They claimed that they had never received my request, but I think they didn't want to deal with us and knew we would go to the FPPB anyway. So we proceeded with our complaint to the FPPB on September 18.

In our submission, we told the Board that people had been unable to enjoy their properties or care for their yards because of the smell. Some of them had even confined themselves to their homes. Others had disconnected fresh air intakes on air conditioning units to attempt to minimize the odour. People without air conditioning and with young children were particularly hard hit.

Here's one example of the situation that we faced. Old Joe's wife invited Dorothy over to pick raspberries. The last several days had been sunny with a light southeast wind. As the sun began to set, the light perfectly showed off their beautiful yard. Because it was surrounded by heavy trees, the smell from the barn would linger until a strong wind cleaned it out. Dorothy went into the garden to pick raspberries and described the smell as lung clenching. She's tough, so she persevered until she had filled her buckets, but I could detect the smell on her clothes when she got home. I can't imagine having to live in those conditions. That week, Old Joe even had to cancel a barbeque he had planned with camping buddies. No one should have to endure that. It's just not right.

When they initially asked for an expansion, the colony had assured the residents that the odour would not be a problem. We'd had several meetings with the colony over the summer, and they even acknowledged that the smell was unacceptable. However, when we reiterated that we were willing to work with them to obtain research money and get the university involved in this project, the colony again declined our offer. That was because they already knew why the smell was so bad, and we were about to find out.

We assured the FPPB that we intended to pursue all avenues at our disposal to have this problem rectified. In our complaint, we had the signatures of all the people who were involved in the Concerned Citizens of Big Island (CCBI). We had testimonials from most of them about the effect the smell was having on their lives. We kept charts on wind direction and smell issues in people's yards surrounding the barn. We had done a thorough job.

On behalf of the Big Island Recreation Club, Harold submitted a second complaint to register a concern about the smell caused by manure spreading on the evening of September 1, 2007. A couple from a local village had rented the hall for a wedding celebration. On Wednesday of that week, the colony had started spreading composted manure on a field less than a mile west of Big Island. The year before, the colony had spread manure on the same weekend on the same field and had been reprimanded by Conservation for not composting it properly. We had reported this latest incident to Conservation, but had not received a reply. It had been a warm evening, and because there was no air conditioning, the doors and windows were open in the hall. The smell wafted into the building, making the situation quite unpleasant. The groom, who didn't understand where the smell was coming from, got annoyed with Harold. When Harold explained it to him, we gained another ally for our group. People started leaving early, and the evening was ruined for everyone, especially the couple. The hall returned the fee they had charged and apologized to them and their guests. If people in the community didn't think that this barn was a detriment before this incident they certainly thought so now. It was disgusting!

***

ON DECEMBER 14, WE RECEIVED THE FPPB investigation report. Among other things that stood out was an admission by the colony to having a 1,150 sow, farrow-to-finish operation going on in the barn. That was 1,437 animal units (AU), far exceeding the 889 they were permitted. This was clearly the reason for the smell from the barn.

I called both the engineer from the province who had written the report and the chair of the FPPB to confirm the information. Both of them said this is what they had been told. This meant that the colony was forty percent over the 889 AUs they had promised us they would respect. I must have phoned half a dozen agencies, including the RCMP, until I found out it was the RM's responsibility to enforce The Planning Act of Manitoba. The province had given the most important function to the least capable government body. I couldn't believe it, and immediately called a meeting of our group. I also called the director of the FPPB to set up a hearing with them.

We asked the RM to meet with us on December 18. After a brief review of our interactions with the colony up to that point, we shared with them the results of the FPPB investigation. The colony's action was an infraction of The Planning Act, and we urged the council to investigate the matter and carry out their responsibilities. We asked for a written response and expressed our wish to be kept informed on all developments in the situation. It seemed to us at the time that the council was listening and a few of them indicated they supported us and were shocked by the behaviour of the colony.

CCBI met again on January 8, 2008. We reviewed our activities from the past year, and discussed the investigation report that stated the colony had an operation of 1,437 AUs. I said I had talked to the Department of Intergovernmental Affairs just before the meeting, and they were going to *advise* the RM to investigate; they said they did not have the power to *instruct* them to count the pigs. Because the RM was the only government body that could pursue legal action, we told the community we'd wait and see what action they chose to take.

We discussed the upcoming hearing with the FPPB that had been scheduled for January 25, and decided to meet again in a week's time to discuss the presentations people wanted to make. The FPPB wanted to know which of the people who had initially signed the complaint would be attending the hearing; we already knew that the colony would be represented by three or four of their members. I gave out the information about procedures, the time and the date and location of the hearing with the FPPB, and asked people to come to a meeting on January 15 so we could finalize our presentation.

Unfortunately, that meeting was postponed because we were not ready and I was still getting communications from the FPPB about the details of the hearing on January 25. On January 18, I got a copy of an email sent from Sam to the FPPB director, indicating that two people from the community who were neither signers of the complaint nor members of the colony would be attending the hearing. The director then asked for briefs from the people who would be presenting from CCBI. On January 21, I told him I would attempt to get the briefs together, and also that I objected to the two members of the community being present. It seemed to us from the guidelines we had been given that only signers of the complaint or the colony or their legal representatives could be present.

The director emailed back that because these two people were residents of Big Island, they should have sufficient personal interest in the subject matter to present information relative to the complaint. He stated that the FPPB was prepared to allow these individuals to participate. I responded that if that was the case then these hearings should be open to any citizen from Big Island who was concerned about the odours from the barn. I indicated to him that we had worked within the parameters explained to us as we had understood them. We were concerned that these hearings would not be held in an atmosphere of impartiality, and that our concerns were not being taken seriously. We asked that this particular issue be addressed before the hearings proceeded. We offered the hall at Big Island as a location for the hearings where anyone from the community could participate. We might even have a chance for the FPPB to experience the smell if the wind was right.

On January 24, the director emailed me back and said that after consideration the board had decided to proceed with the hearings the next day—the original date set for the hearing. I called him to protest, explaining that we needed more time because we were building a case related to the disagreement over who could attend. When I reiterated I wanted that to be resolved *before* the hearing, he simply said they would invite anyone they wanted. Clearly, they weren't following their own guidelines about who could and couldn't attend, but I was not surprised. I was getting a good impression about how arrogant this board was, but what could you expect from people appointed for supporting the governing political party of the day? The FPPB, government officials, and the colony all knew how powerless our group was under the regulations, and treated us accordingly. Had the provincial government not set it up this way, the process could actually have been useful.

That evening I called my MLA who was on vacation in Arizona and explained the situation to him. He was a member of the opposition, and his advice to me was to attend the hearing, rise on a point of order with my concerns, and then walk out. I called an emergency meeting of our group for that evening. I shared all the communications I had received from the director of the FPPB up to that point, and we talked it over. It was clear to us that the Board would favour the colony in this matter. I told the group that I felt like following our MLA's advice. I knew that the hearings would go ahead whether we participated or not, and the outcome seemed pretty certain. I suggested that if we made an issue of the stand the Board was taking, we would have grounds for a complaint to the provincial Ombudsman on the process they followed. Someone asked me what I wanted to do, and I said I wanted to walk out of that hearing. I had had enough of the dictatorial attitude of the director. The battle lines had been drawn, and the group supported me wholeheartedly. I was so proud of them and the team we were building together.

The next morning, I picked up Harold and we headed into the city. We were both dressed in suits and ties and sat in a waiting room with Sam, Peter, and the others waiting for the hearing to begin. We joked around with them while we were waiting. I think we've done that every time we've had a meeting with them, and I doubt that will ever change. Eventually, we were called into the room, introductions were made, and I was asked to speak. I specifically remember the provincial lawyer was late and came in while I was talking.

I ignored the director and spoke directly to the chairperson. "Madame Chair, I rise on a point of order to express our concerns for the events that have unfolded this week. It was our clear understanding from the director and the act governing the actions of the FPPB that only individuals who had signed the complaint, were members of the colony, or legal representatives for either party would be allowed to attend this hearing. The Board's own guidelines and Farm Practices' guidelines state both parties are expected to be present at the hearing and may be represented by legal counsel. We are concerned that the admission of other individuals to the hearing on the part of the FPPB threatens the impartiality of these proceedings and compromises this hearing. We requested a postponement to allow an opportunity to deal with these issues. That request was denied. We believe in the value of this process and find ourselves in the position of having to write to the Minister of Agriculture asking for an investigation of the FPPB's activities in this matter. It is necessary then for us to withdraw from these proceedings until these issues as we have stated can be properly addressed. Thank you. Good day."

And then Harold and I walked out. I wish I had a picture of the collective looks on their faces. The director's face turned ashen. I thought to myself, don't mess with me, buddy!

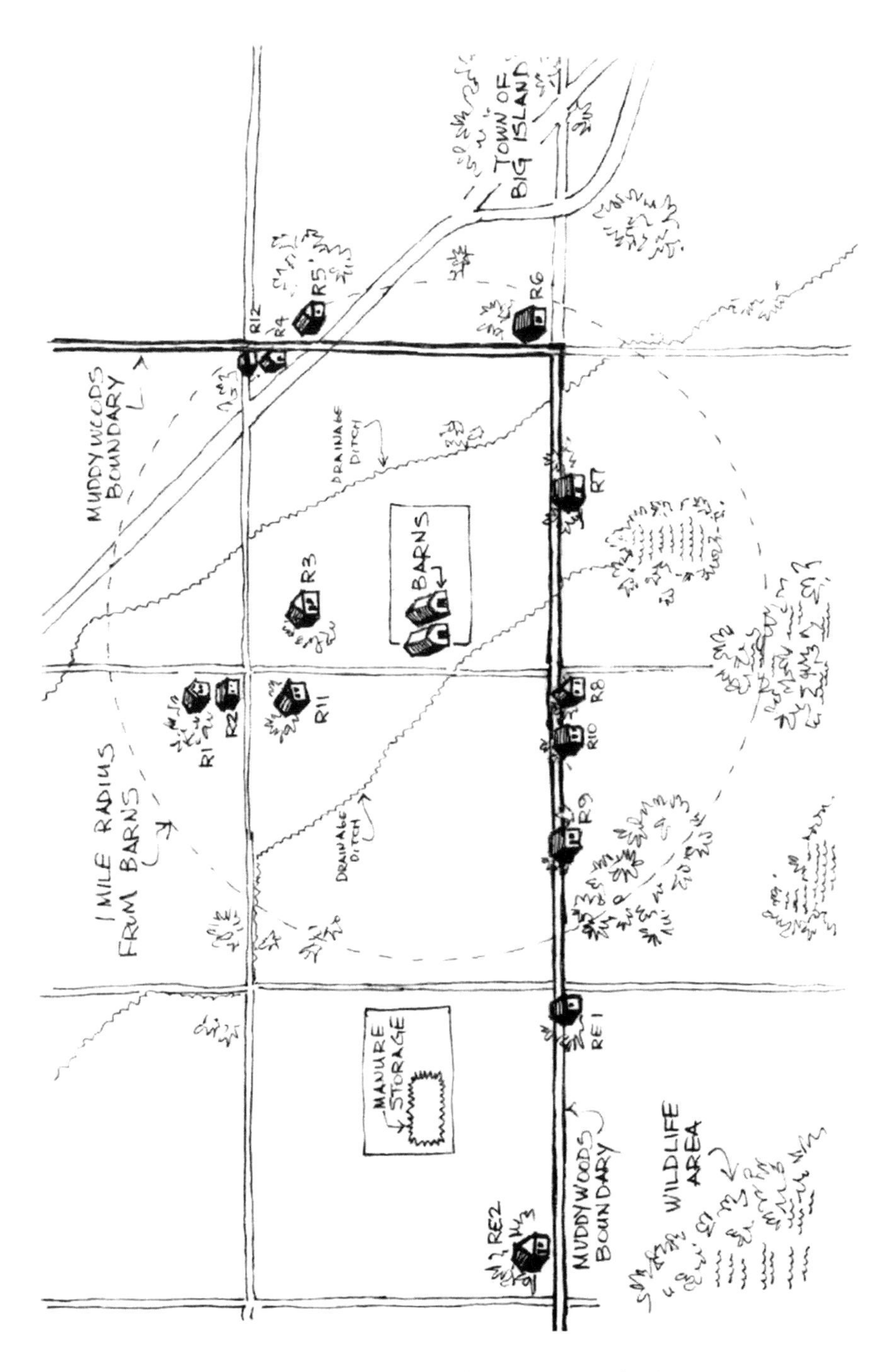

This map shows the residents and the village around the barn. Dorothy and Bill are residence 5, the RM of Greenland borders on the south side of Muddy Woods, and the RM of Snow Valley is on the east. Big Island is in both those RMs. The flow of surface and groundwater is from northwest to southeast. Drawn by local artist Linda Gillies.

CHAPTER TEN

Country Life, Beauty, and Heartache

IT WAS 1955. DEVON TAYLOR WAS IN Grade 1 with me and two others. Four students in Grade 1 was unusual for a one-room school. Devon had a speech impediment, and was pretty hard to understand. We students understood him quite well, but it was harder for adults to make out what he was saying. I was always impressed with how patient Mr. Taylor was with his son. Devon was much nicer than his older brother, who was a bully.

One day, Devon went up to the front of the class and asked the teacher a question. She didn't understand and asked Devon to repeat himself. He was quite used to this, and cheerfully repeated what he had said. By now, the attention of all twenty eight children in the school was focused on this exchange. Both Devon and the teacher became a little embarrassed when she had to ask him for the third time to repeat himself. Devon did so, and, in frustration and embarrassment, the teacher said, "Oh, all right!" The other two Grade 1 students and I got up and started to run outside. Teacher called to us, "Where are you going?" We turned to her and said that Devon had asked if we could go outside to play, and she had said we could.

The other thing I remember about Grade 1 was standing at the blackboard with the rest of the kids and trying to write the numbers from zero to nine. I was the only left-handed child in the group and had difficulty forming the number six. After struggling with this, I started to cry, and Betty Ann Cole, who was in Grade 6, comforted me and helped me make my six. Betty Ann was my hero ever after. Such an understanding and helping hand when I needed it. She let me know I was not alone. She was my first love.

***

I STARTED HAVING MORE TO DO WITH the horses. I remember somebody had come to the farm one day and, for some reason, went into the barn. This woman came out of the barn crying, saying that the manure was so deep under their back hooves, the horses were standing on their heads! Dad never did much of anything around the yard, including cleaning the manure out of the barn, so the horses' back feet were standing on the manure that had built up under them. Mother and I took care of most of the chores and it hadn't dawned on me until then that this could be a problem. Mother, who was always concerned about what other people thought, got quite upset as well. I started using my little wagon to clean up behind the horses, taking some manure every time. I think the horses came to know me and understand that I cared about them.

Mother liked to walk around the yard barefoot, but she had a pair of rubber boots that she wore to go into the barn—they were always left on the ground outside. I remember doing chores with her one evening and hearing a shriek. I turned around and saw a rubber boot arcing through the air. As it reached the top of its arc, it hung momentarily upside down, and a mouse tumbled out. One of those cherished memories.

That fall, we had a great crop of potatoes. I remember Mother and me nearly filling a wagon box. We worked away at this, and somehow managed to store them in a little dugout under the old house. It was a good thing we did because we survived that winter on pork and potatoes. I even remember taking pork and potatoes for lunch to school for six weeks straight! I will admit it did get a little boring, but I seldom complained. I'm still that way. Dorothy wants to throw out leftovers, and I won't let her. It may also have something to do with food security. I don't remember ever being hungry, but there was little variety growing up and my Mother always complained about it. That winter, my father had his nose in a newspaper most of the time.

I remember us going over to the Cole's with the horses one Christmas Eve. We were invited into the living room where Mrs. Cole had decorated a spectacular tree with beautiful bubbling lights on every branch. I was fascinated by the sight of it all. There was warm chocolate for us kids and some small presents, and the adults shared a glass of wine. I remember the feeling of warmth and caring that seemed to fill the room. Mother wasn't used to drinking, and the wine went straight to her head. Eventually we said our goodbyes and wished each other a Merry Christmas.

It was a clear frosty moonlight night. The moon was full and it seemed as bright as daytime. It was biting cold, but fortunately there was no wind. The harnesses were creaking, and the sleigh's runners squeaked over the snow. Mother decided that she would walk behind the sleigh to get over the effects of the wine. She walked in one of the sleigh tracks, which were maybe three inches wide, and every few feet were the imprints of the horses' hooves. It would have been difficult for somebody who was sober to navigate those tracks, and she wasn't sober. My sister and I, and occasionally my dad, looked back and enjoyed the performance. She was more on her side in the snow than she was on her feet in the tracks.

That summer, I was big enough to start handling the horses. We hitched them up to the mower, and I started cutting hay. We always did that kind of work in the afternoon because the grass was drier and easier to cut. The job of pulling the mower was hard enough for the horses, but the horseflies that summer were really bad, and they suffered more than usual. They would sometimes throw their heads back in an attempt to dislodge the biting flies, and swing their tails to try and brush them off. But there was a section of their backs that they couldn't reach, and by the time the afternoon was over, they were quite bloody. Sometimes a horsefly found me and that hurt. Indeed, four hours of that in thirty-degree heat was more than enough for both boy and beast.

Raking hay was far more pleasant. There is nothing nicer than the smell of fresh, cured hay. The horse rake was light and easy to pull. The horses could trot and keep ahead of the flies. We raked the cured hay into long windrows, and when it was ready, we stacked it. I gathered a big bunch of hay under the rake and pulled it up to the stack, Dad pitched the hay up on the stack, and Mother packed it down and shaped the stack. Marlene did what was perhaps the most important job of all—making sandwiches and lemonade, and setting up lunch for all of us in the shade of the stack while the horses munched the new hay. I enjoyed those days a lot.

In wintertime, Dad and I would take the horses and sleigh rack and go out to get hay for the animals. We would use a hay knife to cut the stack, and then load the hay with forks onto the sleigh. As we pulled back into the yard, the cattle would come running and start eating hay off the sides of the sleigh.

Even now when I go to feed my sheep on cold days and open a bale of hay for them, that smell is the most wonderful thing in the dead of winter. I still enjoy it every morning.

***

ONE SPRING DAY, WE ALL DECIDED TO walk up to the drainage ditch a mile north of our farm; Shep came with us. The water was coming down from the escarpment in the Riding Mountain National Park and was boiling under the old bridge across the ditch. There was a big patch of foam where the water swirled around as it went under the bridge. Shep was still pretty young and didn't realize that this was not a snowbank. He stepped out on it and went headfirst into the whirling maelstrom. I ran to the other side of the bridge expecting him to appear. Somehow or other, he got his front paws up on the edge of the bridge, but his back end was being sucked under by the force of the water. Marlene saw this and yelled to Dad who rushed over, grabbed him by the paws, and hauled him out. Shep was quite a furry dog and still had his winter coat. He looked like a drowned rat on the way home.

That spring, I was walking back from seeing my friends late one evening, and there was a beautiful golden sunset as the sun disappeared behind the escarpment. The sky was a bright orange and the mountains looked black in that beautiful prairie scene. Halfway home the road passed through a ten-acre swamp, and as I was watching the sunset, I suddenly saw what looked like a tongue of flame rising from the surface of the water. I was terrified and ran home as fast as I could. Years later, I realized I had seen a will-o'-the-wisp. It seems so beautiful to me now in my memory.

That year, Father joined several neighbours and formed a threshing gang. He bought an old McCormick horse binder on which somebody had put a power shaft so that it could be operated with a tractor. It worked fairly well, but he needed somebody to drive the old G because the controls were at the back of the machine. I was his man! I could drive the tractor standing up, and the steering wheel was right in front of my chest. He fastened a rope to the hand clutch, and when something went wrong, he would simply pull on the rope. The hand clutch had a brake on the pulley. If he pulled hard enough, the tractor would come to a sudden stop, and I would end up against the steering wheel. I never checked, but I suspect that I often had a circular bruise on my chest. Driving that tractor was just the best thing, and I loved it. I would run the mile home after school, and he'd have the tractor already running. He got on the binder behind me because all the controls for the horse binder were at the back. We would work till dark. We were part of that threshing gang until I was about fourteen years old.

***

I THINK IT WAS THE NEXT SUMMER when my mother got sick. I knew things weren't good between her and Dad. She claimed that the lack of money and his refusal to talk to her caused her to have a nervous breakdown. The reason Dad didn't talk to her was because he realized there wasn't anything he would say that would satisfy her, so nothing was better than something. All I know is that she went into the store in Kelwood and broke into tears in front of the storekeeper, who was fortunately a compassionate man. I know, too, that the local Anglican minister talked to her. From what I recollect, she went to bed for probably a week, and I actually saw Dad prepare some meals.

My sister, who was a year older than me, took care of most things in the house. I didn't pay too much attention to what was going on in there because I really didn't care. As long as I got food and could go about my business, I was fine. Looking back, I didn't put much stock in Mother's hysterics either. My sister, on the other hand, was quite affected by it and became scared and worried. She talked to me about it, but from my perspective, it was more of an inconvenience, and I tried not to let it affect my activities outside. I also didn't think too much of it because I had seen my mother in action like this before.

Unfortunately, as a result of all that, I saw a change in Marlene again, and not for the better. As the oldest, I think she took on too much responsibility where her parents were concerned. She didn't have much of a life outside the house, and that kind of emotional turmoil was hard on her. Now it seems clear that she was being affected by the dysfunction in our family. Dad was his usual inscrutable self, and would pay the price for it in the end.

CHAPTER ELEVEN

Failures and Field Pieces

AFTER CHRISTMAS, OLD JOE LEFT IN HIS motor home to spend the winter golfing in Texas with his motor home buddies. Before he left, he asked me to find out what was wrong with his 1950 Ford grain truck. It started well enough, but just wouldn't stay running. We put it in his heated shop and he gave me a key. When I pulled the pan off, pieces of rings and other motor parts fell on the floor. I called Joe and said we would have to pull the engine and rebuild it. He told me to go ahead. I spent part of that winter getting familiar with the Ford 239 flathead V8 motor. Some of the army vehicles Dad drove during the Second World War had that engine. I remembered what he had said about them. He liked the Chevy six cylinder engines better. I guess that's why he bought Chevy cars.

***

ON JANUARY 29, 2008, WE RECEIVED A letter from the rural municipality (RM) of Muddy Woods indicating they were reviewing the situation regarding the colony and clarifying a number of details concerning their operational activities. They also committed to updating our group, the Concerned Citizens of Big Island (CCBI), as their review process continued. Then on February 5, our MLA wrote a letter to the Minister of Agriculture Rosann Wowchuk, suggesting a meeting with her, the

RM and CCBI. We asked her to support our idea for that meeting to take place in Big Island where members of our group could share their concerns.

On February 14, I contacted the Farm Practices Protection Board (FPPB) to ask when the results of the hearing would be available, and was informed by the director on February 25 that the report would be emailed out the next day. They had gone ahead with the hearing without our presence; walking out hadn't done us much good. On February 28, the director contacted me and said they had run into a "speed bump" and didn't know when the FPPB order that was decided at the January 25th meeting would be released. That order considered the stink emanating from the hog barn itself, and the stink caused by the spreading of manure that had ruined the wedding in the community hall back on September 1 the previous year. I never found out the reason for the delay, and they weren't about to tell us.

On March 6, our MLA again wrote to the Minister of Agriculture, as well as the Minister of Conservation, expressing his dismay that we had received no response whatsoever in regards to his previous letter. On March 11, I heard from a staffer in the Minister of Agriculture's office who asked me to write another request to the Minister. On March 13 I called the office and left a message with the receptionist asking them to confirm that they had received my email. On March 19, I phoned the Minister's office again asking to confirm they had received my email and asked the staffer to call. When he finally called, I had had enough. I prided myself on the self-control I had developed over many years of administration, but that day I let fly with both barrels and allowed myself liberties that I told myself they had coming. I could defend myself and give as good as I got when necessary. I worked with that staffer later and neither of us ever mentioned anything about what transpired in that call. However, I noticed that I got quicker responses from the Minister's office after that incident, so I guess I did get my point across—Don't trifle with me!

On March 26, I received the FPPB order, dated February 28. As we expected, our complaint with regards to the odour emanating from the colony's hog barn facility was dismissed. We immediately appealed the decision. To add insult to injury, they kept the fifty-dollar fee we had paid for booking our complaint. We were expecting that to happen as well. However, the complaint from the community hall board was supported, regarding how the colony spread the manure. This was to prove an important development down the road. The FPPB had included a number of recommendations to the colony to modify its future manure application practices. To add further insult to injury, the hall board had *their* fifty-dollar fee refunded.

On March 27, I contacted the RM of Muddy Woods for an update on their investigation of possible violations of The Planning Act. That same day, I contacted the Ombudsman's office. One of our members was in touch with a lawyer that we thought might be helpful in the situation. The next day, I talked with several people in the Ombudsman's office, and they agreed to start an investigation the week of April 7 to 11. On April 3, I started submitting documents to the Ombudsman. The same day, one of our members heard from the Minister's office that we would be receiving a letter regarding our complaint to the FPPB on that day. Another member talked to the reeve of Muddy Woods, who said they were taking firm action with the colony. We were on a roll. No FPPB decision was going to dampen our spirits.

On April 8, we talked to the Ombudsman's office and told them we'd let them know what was happening with the RM, as well as keep them up to date with our correspondence with the Minister of Agriculture. The Ombudsman started their investigation. That day, I received a letter to my March 6th request from Stan Struthers, the Minister of Conservation, who advised me that staff were awaiting additional information regarding the manure storage facility from the engineering firm hired to do an assessment. As a result, he felt it would be premature for him to schedule a meeting with our group. On April 10, I received a letter from the Minister of Agriculture. In that letter,

Wowchuk stated that we had no grounds to complain about the process used by the FPPB, and she would not meet with us. I responded to this with another letter, asking again for a meeting and pointing out the misconceptions in her letter describing the actions of the FPPB. I didn't receive any response. Apparently we didn't have too many friends in that department.

I was at Old Joe's place installing the motor I had rebuilt over the winter. For some reason, we started talking about various ways we could demolish that hog barn. After considering a number of options, we settled on an artillery piece that we could set up on the municipal road. Trouble was, we didn't know what calibre we would need. I told him I would ask around, but, of course, it never went any further than that. We got the motor in the truck, and it ran like a charm. Joe asked me what I wanted for the work, and being a little short of feed for my pigs, I told him to bring over a load of oats in the old truck. One lovely sunny day, he showed up with that beautiful old truck loaded with grain. I can still clearly see him driving on the yard. I was proud of the job we had done on that truck.

On April 10, Young Joe talked to the director of the Manure Management Department of Conservation, who indicated there was trouble with the manure storage facility at the colony. It turns out that the colony had, in fact, entered an agreement with a company to capture the methane gas, but when the colony had started making unengineered modifications to the manure storage facility back in 2006, they were stopped by Conservation from going any further. We were told that this company was considering suing the colony for breach of contract. I will admit we were rooting for that company.

On April 11, I called Conservation and made the point that the colony was in violation of the moratorium that the province had on further hog operation expansions. By increasing their number of pigs on the farm, they had in fact expanded. Unfortunately, that didn't seem to get much traction with the government. Because of those modifications to the manure storage facility, on April 14 I discussed a possible failure of the facility wall with the director of the Manure Management Department of Conservation.

That same day, I met with a reporter from the *Winnipeg Free Press*, and on April 27, we appeared in a photo on the front page, with the full article inside. We stated our dissatisfaction about the provincial and municipal governments and the FPPB and their lack of any action on our behalf. In the same article, the reeve of Muddy Woods complained about us—quite a shift in the RM's position. He suggested that the colony had been raising pigs long before any of us were around. Actually, the colony had been started in 1947, and some of our members, whose families had owned their land for a hundred years, were quite offended. I knew where those remarks were coming from because Sam had suggested at the August 15th meeting the summer before I should go back to the city where I belonged, and I needed to find more to do to keep myself occupied. As an Anglo-Saxon heterosexual feminist colonist white male, I don't recall having experienced discrimination at any time before. If we didn't already know where we stood with the RM and the colony, we definitely knew now.

On April 30, we received a letter from the RM stating that they had given the colony until December 31 to reduce the animal unit (AU) number to 889.1. This was about a year after we had reported them. The letter promised to verify when the colony had achieved that level. Our lawyer we had hired wrote the RM a letter suggesting the agreement should be in writing and registered against the colony's lands by way of a caveat, in order to protect and enforce the adjacent landowners' rights. Within the agreement, the colony should also consent to the municipal inspector entering the premises upon reasonable notice at any time in order to confirm that there were no more than 889.1 AUs in the facility. That makes sense, doesn't it? I thought it was a great idea.

On May 13, we received a notice of public hearing on the application for a Conditional Use Permit on the part of the colony. Their proposal was to allow for a contractor's yard and the establishment of a storage and metal fabrication facility. No clear guidelines were given as to the exact size of the building, type of footings, flooring materials to be used, electrical equipment, plumbing, or ventilation.

We prepared a brief to express our concerns to the RM that the application could not be taken at face value, considering the experience with the hog barn. We did not want this project to affect the environment and our way of life the way the hog barn had. We had also been told that the local planning district, which is responsible for planning and issuing building permits for those RMs that belong to it, didn't want anything to do with Muddy Woods. In our brief, we asked for dust control, a cover on the manure storage facility, hours of operation between 8:00 a.m. and 5:00 p.m., chain-link fencing, control of emissions and pollution, and setback distances from property lines.

We went to the municipal office to collect documentation on the proposal submitted to the RM by the colony for the Conditional Use Hearing scheduled. That evening, representatives from the colony visited a neighbour they thought might be sympathetic to their cause with concerns about our involvement in the hearing. Who had informed the colony that we had applied for this information, and why? We had a pretty good idea it was the council themselves.

We concluded our brief to the RM by saying we were deeply concerned that they were considering another application for development at the colony before sufficient research had been done and issues related to the hog barn resolved. It was our considered opinion, and the advice of our lawyer, that these issues needed to be properly addressed by both the colony and the RM before permission was given for further development. The letter from our lawyer to the RM urged them to enter a development agreement with the colony allowing for semi-annual inspections of the hog facility by one municipal inspector and one designated member of the community.

On May 30, we met with our lawyer and the council at the Conditional Use Hearing. The council informed us that they had already approved the conditional use permit. They had included some provision for dust control, but that was all. Words cannot describe how frustrated, angry, and disappointed we all were. I had the feeling of being subjected to horrible, dirty, and frightening abuse of power such as I had not experienced since my childhood. The lawyer was also upset how things had turned out and gave us a generous break on our fees.

The RM had been given the opportunity to put structures in place that would have dealt with the problem, but they didn't do it. We knew then we were in for the long haul.

***

OLD JOE AND I WERE KEEPING AN eye on the colony and the spreading fields from his small plane. I often flew with him, sightseeing along Lake Manitoba and other places. From the air, we could clearly see, just to the north of us, the remnants of a branch of the Grand Trunk Railway line that had gone bankrupt in the early 1900s. I loved to fly with Joe and I miss him and those flights we took.

On June 2, we noticed a spill site, which indicated that the pipeline to the manure storage facility had ruptured. We took pictures and sent them off to Derek Clarke, the local conservation officer. He went to the site and talked to the colony. When he reported to us, he claimed that the spill was less than ten thousand litres—the level at which they needed to report—and it had not escaped to nearby waterways. The colony was told to clean up the solids in the area because a concentration of nutrients in the hog manure would not allow any vegetation to grow. That meant that if there was overland flooding, nutrients such as phosphorous and nitrogen would be washed downstream into the watershed.

The size of the spill we had seen, however, indicated to us that it exceeded the ten thousand litres of manure threshold. I sent an email to the officer saying we calculated the size of the spill to be 2.2 acres and the spill in excess of twenty five thousand litres. I requested a review of his findings and told him we would be consulting our lawyer in a week's time if we didn't hear from him. Shortly after, I received a phone call from the officer laughingly stating he had missed a conversion on the calculation of spill volume, and it indeed appeared the volume was likely in excess of ten thousand litres. We failed to see the humour.

The colony had reported a spill the previous year and Conservation was now investigating whether this new spill was a related or separate incident. I responded to Clarke that Young Joe and I had met with colony representatives back on April 29, 2007, at which time they had told us of the manure spill that they had reported to Conservation previous to that date. Therefore, I concluded, these two incidents could not be related. Clarke phoned me back and said that we were correct. Not surprisingly, when I asked what consequences the Department was going to place upon the colony, he indicated that they had already done so, with a review of the regulations. In other words, no consequences.

What do you do with that except write about the blatant absence of any enforcement.

CHAPTER TWELVE

Horses, Tractors, and Automobiles

WHEN DAD BOUGHT THE '47 CHEVY, HE gave me the '40. After Mother had her breakdown, she decided she needed to be more independent. There was really no way for her to go anywhere, and she didn't want to drive the tractor or the horses because that didn't look right. She was not getting along very well with Dad and didn't want him to teach her to drive. She could see that I handled the '40 pretty well and told me I was going to teach her to drive the car. So I did.

She was doing well, I thought, but somehow or other, she didn't follow through with getting a licence. There was no test in those days; you could just walk into the municipal office and buy your licence, as I did when I was sixteen. I told her she could only drive my car if she put the gas in it, but I don't think she had any money. Mother had received a small inheritance when her dad died and bought several head of cattle with it, but Dad had sold them when we needed the money. So she was stuck there. I can understand how she must have felt.

The summer of 1959, we started making hay with the Taylors. They needed the help and we needed their equipment. Mr. Taylor had a Styled John Deere B row crop and a New Idea side delivery rake. He knew I had been driving my dad's John Deere G and the '40 Chevy, and told me it was time for me to join the haying crew. He taught me how to operate the outfit and then turned me loose. I loved driving that little John Deere tractor.

I did all of the raking on our place and most of it on Mr. Taylor's farm. There were growing pains, of course, getting the rake tangled up in some old barbed wire that was buried under a blow dirt bank, and making some of the windrows too big for Mr. Taylor's old R and his Case baler. His brother had to come over with his new 720 and Massey baler to do the job. That outfit was brand new and beautiful. I used to own a 720 like that and worked our fields with it. When my neighbour wanted it, he traded me his International 656 tractor for it, and it is still in use today.

Making hay with the Taylors was lots of fun. On their farm, nobody got too excited when equipment broke down and forced work to stop until repairs were made, or worry that the weather might break and the job would not get done. Not the Taylors. They would sit down and have a smoke, consider the problem, maybe go for coffee, and then fix it after it had been thoroughly discussed. They seemed to have infinite patience with this old equipment, and took everything else in stride.

I still remember the Case baler. I'm not sure how many bales it actually tied. Sometimes, when it was being particularly obstinate, Mr. Taylor would sit on the back and tie the bales himself while another person drove the tractor. We always pulled a wagon behind the baler and stacked the bales on it. All up, there were two trailers and two tractors, sometimes including the little B that I drove, and four men—one driving the tractor pulling the baler, one stacking the bales on the trailer, and two hauling the loaded rack into the yard to unload the bales.

Mr. Taylor used to like to throw bales himself and often built the stack. I remember once seeing him walk backwards along the top of the stack and accidentally step off the end. He did a complete somersault in the air and landed on his feet, still holding the bale! We rushed over to check on him, but he didn't seem to be hurt. He wasn't the smartest farmer in the world, but he was definitely the most entertaining. He had the biggest grin, and when you looked at him, you knew life was good. I still smile when I think of him.

***

WHEN FALL CAME, IT WAS THE ANNUAL threshing time again. Dad had bought a combine when he started farming, but it wasn't very good. He and I cut the crop with the old G and the binder. My sister and I helped him stook, standing up the sheaves in little bunches. The binder cut and tied the grain into bundles called sheaves, and left them in rows on the field. We would run along the rows and set the sheaves up into stooks.

When we started threshing with the Carters and the Smiths, we would all gather at one of our farms with teams and racks and the threshing machine. We would work together until everyone had their crop in. Usually the racks were pulled by a team of horses, but Mr. Carter was short of a team, and he owned a little BR John Deere. I was still too small to handle a fork and pitch sheaves, but I could drive the tractor, so that was the job I got. I started off in too high a gear and dumped part of a load. Mr. Smith wasn't happy with me, and I apologized, but I knew the problem was that he hadn't built the load very well. I kept those thoughts to myself, however.

When Mr. Carter had pulled the threshing machine up to a little granary, he needed someone to go inside to shovel the grain to the back. I volunteered, but quickly discovered it was very dusty in there. My face was soon black from the grain dust, but I stuck it out. Even Dad was impressed! I coughed and spat up black globs for about two days.

On the plus side, I was getting quite the reputation in the community. If I could possibly do a job, I would do it, no matter how difficult. There were five or six of us about the same age that were available to work in the district, but people wanted me if I wasn't already spoken for.

The morning of October 7, 1959, we awoke to about two-and-a-half feet of wet, heavy snow. Nothing could move except cattle and horses. About nine o'clock in the morning, we saw a black figure crossing the section to the southwest of us. It was Harry, our neighbour, who lived kitty-corner on that section a little more than a mile away. We watched him come across, lifting his feet high in the air to take each step. He got to our place at about one o'clock in the afternoon. He was exhausted, but he was also desperate and had come for help. We'd never had a phone on the farm; we didn't have electricity, either, for the first five years. Mother made him coffee and gave him lunch, and Dad went out and harnessed up Trigger and Buttermilk. Harry had a flock of about 350 sheep and no horses. His sheep were stranded in the pasture and couldn't move because of the snow.

Harry and Dad had to put the rack on the sleigh that afternoon. The rack was still on the steel trucks because no one had expected this early snow. When that was done, Harry, Dad, Shep, and I climbed up on the rack and headed out across the section. The horses were plunging through the snow, and Dad stopped every hundred yards to rest them. It was about three o'clock in the afternoon when we got to Harry's farm. We loaded a small bundle of hay on the rack and headed out into the sheep pasture. It took us an hour to get out there and I could immediately see why Harry needed our help. The sheep had packed down a little area in the middle of the pasture and were huddled closely together. We threw a little hay down to them, but not too much because we were worried they might trample each other.

Dad got the team and sleigh turned around and started back towards Harry's yard. Shep and I got on the ground behind the flock, and Harry threw a little hay off the rack to entice the sheep to follow. Going back, we made quite the procession, which was probably a quarter of a mile long. Trigger and Buttermilk pulled the sleigh while following the tracks back to the yard. Dad was driving, and Harry was throwing off little bits of hay to encourage the flock along. Shep and I followed behind the sheep to keep them moving. Eventually the procession arrived at the barn. Our job was done, and Harry could take it from there. Dad, Shep, and I headed for home along the trail we had made earlier.

That winter, we made a few trips to town with the horses and sleigh because most of the roads were impassable. With the help of Trigger and Buttermilk, Dad eventually got the '47 out to the highway, where it sat for the winter. We would walk the mile, hoping the car would start when we all got there. Going to Kelwood was usually a great adventure, and we were quite disappointed when the car wouldn't start, forcing us to walk back home again.

There was a lot of crop left out when the snow came. Those of us who were still threshing did fine. People just put their racks on sleighs, and on a nice day in winter, they would be out with their horses and sleighs, gathering sheaves out of the snow, and pulling them into the yard where they would thresh with their tractor and threshing machine. I still remember those bright sunny days in February, walking through the snow over to the threshing gang and pitching in.

***

THAT YEAR, I HAD ASKED THE TEACHER if I could sit at the front of the class. When she asked why, the reason I gave was that I could better concentrate on her lessons. She liked that, and I gained a little favour with her. It did help me concentrate better, but that wasn't the reason I wanted to sit at the front.

Among the other jobs we hadn't got to that fall was neutering the young male pigs. We had left it pretty late, and the pigs had grown fairly big, making the job difficult, so the Taylors came over to help us. When we had finished with the pigs, one of the Taylors couldn't find their mitts. I thought I saw it, and being helpful, rushed over to pick it up. What I saw, however, turned out to be one of the pig's testicles that had been thrown out on the snow for the dogs. Mr. Carter's brother saw this, and then the word was out—I couldn't see very well.

I felt embarrassed and ashamed of my disability, and was devastated that people knew. My parents did not seem to be concerned about it. There was nothing unusual about their lack of concern for me. I think it was the same for Marlene.

It was also that winter I started riding the horses bareback. I began with old Trigger because he was quieter than the mare. I rode him over to my friends, and we would play in the field with him, pulling a toboggan. He didn't want to gallop even though I kicked him mightily as I urged him on. The only time I could get him to gallop was on the way home when we were in sight of the barn.

Eventually I started riding the mare. I was careful at first and didn't go too far with her. She could run like the wind and loved it. Things really got good when our neighbour, Harry, lent me an old saddle. His family had emigrated from Holland in the '30s, and brought along a saddle without a horn. Lucky for me, it did have an open iron loop in front that worked just fine for tying a rope attached to a toboggan.

It's exhilarating to fly along on the back of such a powerful animal, and that saddle and the mare gave me a whole new opportunity. We always saw westerns in the theatre in Kelwood and I thought I fitted right in. It felt great.

One of the neighbours told my dad that I was wearing out those horses, but Dad said no. He figured they would be in great shape to work for him in the spring.

Aerial photograph of the farm in 1960.

CHAPTER THIRTEEN

Drainage Disasters

WHEN THE HUTTERITE COLONY STARTED CONSTRUCTION ON the metal fabrication building, they began by creating an excavation just west of the colony building site and south of Young Joe's property. In May, before the project was approved by the rural municipality (RM) of Muddy Woods, they began hauling material from a site just west of the colony. Because we had no idea what they were up to now, Old Joe and I went to RM with our concerns that this excavation would be a second manure storage facility. We could see them hauling the material across a municipal road with heavy construction equipment and creating deep ruts. We were concerned that the pit was quite deep and could compromise the aquifer. We also worried that the spillage from the pit would leak into the nearby municipal ditch, which drained into the Omand's Creek watershed.

It was apparent at the meeting that the council was unaware of the excavation. So we asked if they took any steps with Water Stewardship to ensure the integrity of the aquifer. Had the RM accepted a fee for the approval of the Conditional Use Permit without requiring a detailed plan and researching possible environmental and other consequences from this construction? Did they know exactly where the pit was located and its proximity to the drain? As expected, we received blank looks from the Muddy Woods councillors. There had been little or no due diligence that we could see. Guilty on all counts.

Dorothy and I kept two cattle beasts on the farm, butchered one every year, and bought a dairy calf to replace it. We always had two animals on the farm. One day, our heifer got out and disappeared. We went over to Old Joe's place, and he took Dorothy up in the plane to look for her. They finally spotted her tracks leading into Joe's bush. They landed, and we found her hiding, holding her breath, in the bush right beside the hangar. Dorothy certainly enjoyed going up with old Joe on that beautiful summer evening. He overflew our place and it was the first time she had seen it from the air.

***

ON OCTOBER 1, WE AGAIN COMPLAINED TO Conservation about the smells coming from the colony. There's quite a difference between the smell of hog manure and chicken manure. The chicken smell is more acidic and can cause the sinuses and eyes to water. The colony disposes of dead animals in with the chicken manure in a composting bunker. Sometimes those animals are not completely decomposed when the bunker is emptied and spread on the fields. It was the colony's practice to empty the bunker filled with this mixture about that time each year. The previous year it had spoiled the wedding in Big Island.

Derek phoned me and said that he had inspected the field and didn't see any uncomposted animal remains. I asked him if he had walked out into the field to look closely, and he told me that he had looked with binoculars from the road. I suggested to him that he should have gone out in the field to have a closer look, and he told me he didn't because of bio-health concerns. I replied, "Your bio-health concerns? I can lend you a pair of rubber boots." When he said he was already leaving the area, I suggested to him that all of the people who lived within a mile of this field also had bio-health concerns, but unlike him, we couldn't just leave and dismiss it. I was not impressed.

I also contacted the Farm Practices Protection Board (FPPB) because they had written an order about this problem the previous year. The order was violated in that no neighbours were consulted, the smell was a problem over a weekend, the manure was not incorporated after it was spread, and the field was close to the community. I suggested that if the inspector was serious about finding out whether the colony was following the terms of the order, that person could talk to us.

Not surprisingly, the inspector never called. Instead, I received an email report on October 22 stating that the terms of the order had been discussed with the colony and no other action had been taken. I then asked the inspector to attend a meeting at Big Island with the eleven families directly affected by the odour, but my invitation was declined.

After an article appeared in the *Winnipeg Free Press* on October 26, 2009, entitled "Won't back down from hog-farm fight," Harold came to see me. He had always been uncomfortable with our use of the media to criticize the RM. Harold had been a salesman and was very good at establishing relationships with the municipal councillors. He was also quite successful at getting grants for his various projects around Big Island.

When Muddy Woods began to threaten that, because of the negative publicity, they would withhold support for projects that benefited our community, Harold asked me to stop using the media in our struggle. I refused, saying that the media was about the only tool we had because there was no recourse for us through government channels. He didn't agree with my position, and after that, Harold abandoned our group. He was well known and respected by the councils and the colony, and his departure from the group hurt us. And because he was the one who had convinced me to lead this struggle—in spite of my misgivings and Dorothy's objections—I felt quite betrayed. His sister who is part of the group believed he would come back, but he never has. I thought the grants from Muddy Woods didn't amount to that much anyway. I guess everyone has their priorities.

And so the year drew to a close. It had been a busy one, and we were left with many questions still unanswered about the hog barn and the new metal fabrication plant.

***

ON MARCH 12, 2009, WE RECEIVED A letter from the RM of Muddy Woods. It included a report by Manitoba Agriculture that indicated the colony had reduced their herd in the hog barn to 852 animal units (AU), fourteen months after we had reported the infraction. We did think the odour had improved a lot.

On May 4, Young Joe called. He had noticed a two-foot drop in the level of water in the borrow pit next door to him. Our first thought was that the water had gone down and contaminated the aquifer. We called Water Stewardship and told the senior Water Resources Officer Geoff Reimer, who was in charge of water control works and drainage licensing, about our concern. By May 24, when I hadn't heard anything, I emailed again. On the 26th, I got a response.

Reimer explained that the colony had installed an eight-inch pipe at the southwest corner to act as an overflow outlet. That outlet drained into ditches that empty into Omand's Creek. He went on to say that the bedrock in that area was thirty-five to fifty feet or more below the surface of the soil, and in his opinion, infiltration rates were negligible through this type of material, which was mostly clay. After installing the pipe, the colony was allowed to apply for a water rights licence, and no enforcement was taken. Surprise, surprise.

Young Joe was livid with this response. He had been attempting to obtain a water rights licence for a project that he had wanted to do on his farm for more than a year. He emailed Reimer with his frustration about their findings. Joe also suggested that based on their findings re the colony, he could proceed with any surface drainage project and, if requested by the Water Resources Department, then subsequently apply for a licence without penalty. This precipitated quite the exchange.

Reimer's angry response indicated that he felt his department had done more than enough to deal with the colony's actions. He also implied that we were deliberately targeting the colony with all our enquiries, which made me see red! Then he suggested that Young Joe had drainage projects on his land for which there were no water rights licences, and as a result, Joe could be charged for violations. The drainage projects he was referring to at Joe's had been in place for more than twenty years, long before the current legislation even existed.

I had been copied with all this correspondence, so I replied that I was unclear as to what the intentions were behind some of Reimer's statements, and asked him to clarify those for me. I reminded him that he and I had never had a conversation about our group's concerns with the colony. I also asked him if the problem had become our group reporting these infractions, not the infractions themselves. I included a list of infractions that had occurred so far without any penalties being imposed, and asked what incentives there were for the colony to follow the regulations. I suggested that the purpose of his department was to enforce those regulations, and ultimately to protect the community and the environment from inappropriate practices.

I concluded by saying that I wondered why he had mentioned drainage works on other properties, such as Joe's. I mean, if those works were problematic, wouldn't it be the responsibility of his department to deal with the situation? I asked for contact information for his immediate superior and said I was looking forward to hearing from him. When I received that information, I complained to the supervisor about Reimer threatening Joe, and waited for a response. In due course, I heard from him, and he was not, shall I say, sympathetic. We had no further dealings with Reimer again. He moved on. This seemed to be the way Water Stewardship dealt with our concerns.

***

WHEN WE HAD RECEIVED THE COLONY'S MANURE Management Plan (MMP) back in 2004, these plans were public information, but shortly after that, the government shielded them from the public through the regulations of the Freedom of Information and Protection of Privacy Act. On January 9, 2008, we had requested the MMP for July 2007. When we were refused that information, we appealed to the Ombudsman's office. In May 2009, we finally received the MMP, but only because the colony had written about their new barn in a hog producer magazine, and the information was determined to be in the public realm anyway. However, the Ombudsman advised us that this did not mean we would get that information every time we applied, only in this instance.

In early June, I realized that I still hadn't heard anything from Conservation about the unlicensed human waste lagoon that had been discovered on the colony in 2004. When I contacted environmental officer Mike Baert about the situation, he asked, "Is there a problem with the human waste lagoon?" I responded that the last we heard was that the colony was to have an engineering report done on that lagoon. He indicated that there wasn't any documentation in their office about such a report related to the lagoon, but they were in fact meeting with the colony the following Wednesday to go over all environmental concerns.

On July 2, we reported another spill at the pipeline leading to the manure storage facility to yet another officer at Conservation, Donna Smiley. Nothing seemed to happen with that complaint either, and in fact we never even got a response. It appeared they were cycling their officers through our mill and this one apparently also got lost in the shuffle. No one seemed to want to deal with us. I can't imagine why.

Given the help we had been receiving from our lawyer, we decided to ask the community for a contribution to pay the lawyer's fees. On July 13, I made a report to the Community Hall Board, gave them a rundown of events that had occurred until the present, and asked for their support and a financial contribution. They were very generous financially and forthcoming in their support for our efforts. I appreciated their actions very much.

On July 15, we met again with our lawyer, and learned that unfortunately, the prospects for legal action did not seem to be that great. We talked about the use of the press to shame the colony into appropriate action, and a few of our group felt this might make the situation worse with the RM and create negative publicity for Big Island. I realized Harold was still trying to influence some of the members of our group, the Concerned Citizens of Big Island (CCBI).

On July 21, we went to see Muddy Woods Council. We thanked them for organizing the AU count. CCBI was in agreement that there had been a substantial reduction in odour since the number of AUs had been reduced to 852. We stated our concerns that when the hog prices would go up, so would the number of pigs at the colony. We told them there had been no open and transparent or even honest communication from either the colony or the RM to our group and the community. We pointed out that the building of the barn and the manure storage facility had been opposed by both surrounding RMs. We reminded them that the colony had promised they would take care of any odour issues that might develop. We said we had reported eight violations that had been investigated, but to our knowledge, not a single penalty had been imposed. We mentioned that every time pressure was brought to bear on the colony regarding improving their operation with the hog barn and lagoon, the issue of money came up, and yet they had constructed a massive new grain handling facility, bought more equipment for the metal industries building, and put up another building. We asked the RM to take a more proactive position on matters with the colony and require three counts per year to ensure the colony was in compliance. We suggested we would launch legal action if the colony exceeded their numbers again, and advised them we would be contacting the other two RMs that were impacted by this operation. We had covered all the bases.

Shortly after that meeting, Dorothy and I attended a funeral at the Big Island Community Hall. When we arrived at the reception, I noticed there was a table at the back of the room with a number of people from the colony, including Sam. I walked straight across towards them. Sam watched me approach, with an exhausted look on his face. The other people at the table looked at me like I was an alien from outer space.

"How are you doing, Sam?" I asked.

He looked at me and said, "I'd be a lot better if you were a quitter!"

I had to laugh. "That's not likely to happen any time soon, Sam."

How prophetic those words were.

CHAPTER FOURTEEN

When Your Blood Runs Cold

IN THE SPRING OF 1960, THE TAYLORS started building a new house. Dad was over there quite a bit helping them. I believe they were paying him for that job, and we were getting some meat from them as well. We were living off that and out of Mother's garden, and occasionally, when we would buy a loaf of white bread, my sister and I thought it was candy. I didn't help much on the Taylor house, but Dad and I did some work in the basement of the new school. Because he was an amazing mechanic and good at all sorts of repairing, people asked him to fix things that no one else could.

When he and I were out at the neighbours' together, he seemed to change his behaviour, like I had seen when we were at his sister's place at Christmas. He was reasonably nice to me when other people were around, but when we were working together alone, it was a different story. He never gave me much direction as to how to do things. I mostly learned when he got angry at me for making a mistake.

As time went on, I got used to the idea that my dad was a different person when we were by ourselves than he was when other people were around. He was careful how he behaved in front of others, which explains why everybody liked him and thought he was outgoing and helpful. He even complained about me to others—while I was in the room!—and it wasn't until I went out and started working for them that they got a truer picture of the kind of person I was. You would think that he would be proud of me, but far from it, I'm afraid. I think I was a threat to him, and I believe Mother had a lot to do with that.

Somehow or other, he got a crop in that spring, but that was it. The local fuel dealer had let him run up a tab, and either he got cut off or didn't want to ask the dealer for more credit, but all summer the G sat hooked to the one-way in the summer fallow; it had run out of gas, the ragweed growing up around it. Dad was never home that year. He was over at the Taylor's working on their house.

That summer, we made hay with the Taylors again; they provided the equipment and the fuel. I worked for them, and in return, they put up our hay. I didn't get paid, but that was not important to me. I didn't mind a bit. What I liked was the respect I got for working. I got to drive Mr. Taylor's little B John Deere while mowing and raking and went there for Mrs. Taylor's amazing lunches.

We had only a few animals on our farm—the horses, a couple head of cattle, some chickens, and pigs. We butchered one of the pigs for meat, and shared it with Harry. When that was gone, we butchered a ewe at Harry's place and shared that. We didn't kill our best animals for food; those we sold. The ewe we killed was unable to bear lambs any more. I learned to like mutton, and prefer it to lamb to this day.

***

THE YEAR THAT WE MOVED THE SCHOOLHOUSE to the farm, Dad began tearing down walls in the old house. He used that material to put a partition across the main room. This divided the space into two large rooms—a kitchen area and the living room area. He built a small cubicle in the living room for Marlene to have her own bedroom. They pushed the organ up against the wall of her bedroom and put a "Toronto" couch behind it. A Toronto couch was a small single couch that could be made into a double bed. That was my bedroom. The partitions he made went up about eight feet. The school had a vaulted ceiling, and he had intended to lower it to meet the partitions. But Dad never seemed too interested in finishing anything at home, so the ceiling was never lowered. Our parents slept on the couch in the living room, and Mother folded it up during the day. Essentially, we all slept in the same room. There was no privacy, so I heard everything.

It was definitely much warmer than the old house, especially when we finally got electricity to the farm. A line was put in that went right past us, so we were able to have it hooked up. In the days before we had electricity, I remember all of us sitting at the kitchen table, with four books pushed up to the coal oil lamp and the four of us reading away.

Around that time my sister started to attend the Pentecostal Church in Glenella. Someone would come and pick her up every Sunday, and I remember how happy she was to be going there. That was her escape. She tried to persuade me to come, too, but I was having no part of it. For a while we had gone to the Anglican Church in Kelwood after Mother's breakdown, but that only lasted about a year.

The Gideons came to the school and handed out New Testaments when I was in Grade 5. There was a schedule at the front of the book. If you read certain chapters every day, you could read the whole thing in two years. I read it through twice. The second time was a bit of a struggle, but even then, I would do my best to finish whatever I started. That was enough religion for me.

That summer, I started working for Mr. Carter and actually got paid; that was nice, but it didn't matter all that much to me because I was quite used to being without money. Mr. Carter had this beautiful little BR John Deere Unstyled tractor, and I mowed hay with it and a New Idea mower. That mower was interesting to work with, and was quite a different design than other mowers. It worked well enough, but was more complicated to fix when it broke down. Mr. Carter was not very mechanically inclined and appreciated me when I fixed it for him several times.

Mrs. Carter was a comfortable woman who loved to cook, and did she cook well. Judging by how warmly he always greeted me, I'm certain Mr. Carter liked me. He had two sons just a little older than I was, and had the gentleness of an older man whenever he dealt with me. He had married late in life and must have been twenty years older than Mrs. Carter. I was beginning to realize that other families were not like ours, and other fathers were not like mine. I quite enjoyed my time away from home and interacting with the men without my father around.

I could ride my bike or one of the horses over to Mr. Carter's to work. I spent quite a bit of time there that summer, earning a little money by cutting and raking hay for them. One weekend when I was over there, a relative came by who was a photographer. He took pictures of Mr. Carter's sons on their riding horse, and then asked me to get on the horse so he could take my picture as well. Several months later, I received a beautiful photograph of Trixie and me. It's one of my prized possessions. I am about twelve years old in that picture.

Mr. Cole was sick that summer, so I went over there and helped them crush grain for their animals. My friend Andrew was a little too young to help his dad with some of that work, but was regularly driving the tractor for him. Mr. Cole would have liked me to work there a lot more, but I was committed to helping Mr. Carter and Mr. Taylor. I also had the chores at home and helped my mother in the garden, so I was very busy. I still wanted to play, though, so after we'd finished crushing, Andrew and I played with his Meccano set for a while.

About this time, we learned that our school was getting a new teacher. I was a little nervous because I hadn't had a good experience with the last one. When I walked into the school, I couldn't believe my eyes. Here was a beautiful young woman, probably about nineteen years old, smiling at me. I fell in love immediately. I was her best student, I was the best behaved, and I got the highest marks in the school.

After about a month, she called me up to her desk and asked me what I was interested in learning. We had just started studying the history of ancient Egypt, and I was very excited about that. I think she saw potential in me, and offered to bring books on ancient history for me to read. I was delighted and told her I would like that very much.

I love history to this day and read it extensively. I would like to have been ten years older when I met her.

My dad had started talking about his war experiences when I was old enough to understand them. My mother hated those stories, so he told them to me when we were alone together. He had enlisted in 1941 in the Fourth Armoured Division and landed on the beaches of Normandy in June of '44. He rode a motorcycle and guided truck convoys in France and Holland. This was history to me, and I was fascinated. I often wonder now if he was telling me these things so I could understand his behaviour better, as clues to why he treated me and our family the way he did. I believe Mother hated his stories for that very reason. It was only years later that I heard the term "post-traumatic stress disorder." It certainly fit my dad.

I knew from an early age that Dad had been engaged to be married when he went overseas. When he came back, the girl didn't want him. He had changed too much. Marlene and I heard that from our mother, who knew about all of it, and I think was threatened by it. She certainly brought it up often enough in a bitter and sarcastic way. I could see he was bothered by her even mentioning it. She used it when she really wanted to get to him, which seemed to be fairly often.

Mom had met Dad at a dance in Brandon and they were married shortly afterward. Things had not gone well from the beginning, and she was an unhappy woman. Dad had no use for her family, and Dad's family had no use for Mother. But my mother was a survivor and had learned to look after herself first. She and Dad had not much going with each other.

One night, I overheard her speak to him softly. "Where did we go wrong with Billy and Marlene? Perhaps we should have another child." My blood ran cold, and then I got angry. I told no one what I had heard until after Dad died, and then I only told Marlene. Mother trusted no one, including my sister and me. Marlene and I had already begun to look outside the home for the affection and acceptance our parents couldn't give us. I think Mother was threatened by that, too, and having another child suggested she was hedging her bets. You see, it was all about her and her needs. Somebody to look after her in her old age. She had given up on Marlene and me, or perhaps even discarded us.

Bill at age 12 on Trixie, the Carter's riding horse.

CHAPTER FIFTEEN

Insults and Achievements

ON TUESDAY AUGUST 4, 2009, I GOT an email from Conservation saying that the files on the colony's human waste lagoon had been found. The last document in the file had come from an engineering firm that told Conservation they had been retained to provide an engineering report for the lagoon. The colony had been instructed to contact the engineer and find out the status of the report. Conservation explained that the time lapse between 2006 and 2009 was due to a complete turnover of staff in that office. No one had followed up on this file, and it had been lost until I asked about it. I know the colony loves me for this one.

Not surprisingly, the engineer's report said the lagoon was unsatisfactory, and when I asked if the colony would have to replace their lagoon, Conservation responded that they would be required to submit their application to build a new one to Conservation's Assessment and Licensing Branch. I couldn't help but think that if the regulations around hog production were as well enforced as those around human waste, we would have been able to deal with this problem long ago. It's ironic that hog manure, which can affect human health and well-being and is full of antibiotics, doesn't seem to be much of a concern to the government. And I don't think I'm wrong about that.

In his article "Air Quality and Community Health Impact of Animal Manure Management," Siduo Zhang of the School of Population and

Public Health at the University of British Columbia says that air pollution from animal manure may pose a health threat to workers and community residents.[3] These exposures have been associated with respiratory and cardiovascular effects, impacts on psychological well-being, and even acute poisoning or death. So I guess I'm not wrong.

***

ON SEPTEMBER 25, WE RECEIVED A COPY of the Manure Management Plan (MMP) the colony had submitted in July. Of most interest to us was the number of pigs that they reported in this plan. The number 911 animal units (AU) was given, but only for the sows, farrow-to-nursery. Compliance was 889 AU, and the rural municipality (RM) of Muddy Woods would allow the colony to be over by ten percent to a total of 977 AUs. They also indicated there were growers (not quite ready for market) and finishers (ready for market), but did not list the number of AUs. If those animals were taken into consideration, the colony had 1,312 AUs in their operation. This was good information for us, and we attempted to obtain the AU count from their MMP on an ongoing basis.

Unfortunately, we seldom got the numbers after that because Conservation redacted the copies. Nevertheless, I kept asking for the numbers until early 2015 when I learned from Conservation that there were no penalties for misrepresenting those numbers on the MMP anyway. Even if we did get copies, the colony could put false information on the MMP without any consequences. Why does Conservation do this?

We talked to our lawyer about our options. It appeared that we couldn't go to the FPPB about the number of AUs in the operation because that was not in their jurisdiction. But before we could take legal action, we still had to go through that process. Even if we did those things and went to court, there did not seem to be much likelihood of

3 Siduo Zhang, "Air Quality and Community Health Impact of Animal Manure Management," *National Collaborating Centre for Environmental Health*, September 2011, https://www.ncceh.ca/sites/default/files/Air_Quality_and_Animal_Manure_Sept_2011.pdf.

success, and we would incur a great deal of expense. We seemed to be at a dead end; it looked to us that the province had apparently sewn that up quite well.

On October 18, we reported a very strong smell from the hog manure storage facility. I phoned the Farm Practices Protection Board (FPPB), and the colony was instructed to pump off the rainwater that had collected on the cover. The cover is heavy, but because the edges aren't a tight fit, too much rainwater on top can weigh it down. That creates an opportunity for the effluent underneath to leak out and onto the cover. The combination of effluent and rainwater was clearly causing the stink.

We went back to the *Winnipeg Free Press* and an article appeared on page A5 of the issue dated October 26, 2009, outlining our grievances and the unfairness and inequality with which we were being treated. The reeve made a number of remarks in that article that made it clear his support was wholeheartedly behind the colony. As usual, we responded with a letter to the editor section for publication.

When speaking with our lawyer, I brought up the subject of the changes in the number of AUs on the colony in 2004 and 2005. In that period of time, we had three numbers—658 on their MMP, 786.8 on their proposal for expansion in 2004 prepared by DGH Engineering, and 888.9 in their request to replace existing facilities. He suggested I write to the RM and request any information, permit, bylaw or resolution on the part of the RM that reflected these changes.

And then on December 31, we received notification from the Ombudsman's office that our complaint—that we had been unfairly treated by the FPPB—had not been upheld. However, that office initiated discussions with the FPPB in regards to the enhancement of the FPPB's operational guidelines. At least that was something.

About that time the director of the FPPB retired.

***

SEVERAL YEARS EARLIER DOROTHY HAD INJURED HER knee when we were moving some sheep. It continued to bother her, so she went for surgery in November 2009 and was on crutches for several months. Just before Christmas, one of Peter's sisters called her to come over to the colony. The colony had received a truckload of coats that had been damaged in a warehouse fire, and laid the coats over several acres of a snow-covered field. They had let the coats air out and selected the ones they wanted for themselves; Peter's sister wanted the rest to go to charity. I suggested to Dorothy she get in touch with the Winnipeg Fire Department. They had a Christmas coat drive every year, and were very happy to take them. Dorothy was on crutches, so she asked me to bring our truck over to get the coats. We loaded the truck right full that evening.

I knew I wasn't welcome at the colony, so when Dorothy was invited in to get some boxes of food, I stayed by the door. The women then brought the boxes up from the basement to give Dorothy some baking, as they often so generously did.

There were a dozen women standing around when suddenly Sam's wife saw me at the door and shouted, "Is that Bill Massey? Why did you bring him here? He's making our name black!"

There was a stunned silence in the room. Dorothy responded that I had come because she was on crutches, and I was helping her to carry the coats. On the way out, Peter's sister indicated Sam's wife and quietly said to Dorothy, "She's gutsy!" That was not a compliment.

After that incident, Dorothy was never again invited back to the colony.

In the summer of 2019, Peter's sister phoned her and asked for a ride to a funeral. She had no other way of getting there. Dorothy obliged. In the past that might have resulted in an invitation to come in for coffee, but not this time. When Dorothy came home, bringing some food from the colony, she told me she was saddened that her hopes of having friendships at the colony had not developed as she had wished they would. The problems at the colony were their own doing, certainly not Dorothy's fault, but she paid the price anyway.

***

ON FEBRUARY 5, 2010, WE LAUNCHED A complaint with the Ombudsman against the RM of Muddy Woods. We had requested information, permits, bylaws, and resolutions regarding discrepancies in the number of AUs of the colony. We were not permitted to access this information at the RM office ourselves, and it had not been provided to us. We had received correspondence from the RM that did not answer any of our questions.

In March, I attended a hog industry conference in Brandon. When I attended these things, I would always identify myself and state I was the leader of a group protesting an illegal hog operation at Big Island. Most people in the industry in Manitoba knew about me by this time, and that approach led to a number of interesting conversations with many different producers and companies that sold goods and services to the industry.

I was in the display hall when I realized that the company that had built the barn at the colony had a booth. I went up to the representative, introduced myself in the usual way, and said my neighbour and I had a question about the barn. Fire away, he said, not knowing how appropriate that phrase was going to be under the circumstances.

I asked if he knew what calibre of field piece would be required to knock the barn down from the municipal road about 200 yards away. His eyes got wide, he realized I was joking, quickly recovered, and then responded in the same vein. "I've got bad news for you. We built these buildings for the American military in Iraq." I couldn't wait to tell Old Joe. He was tickled by the story.

There was a change in directors at the Manure Management Program in Conservation. I wrote a letter to the new Minister of Conservation, Bill Blaikie, expressing our concerns about how Conservation had handled the colony's file so far. I was then contacted by the new director, Don Labossiere, and he and the Director of Environmental Service, Mike Gilbertson, agreed to meet with us. That happened on Wednesday November 3, and it was an excellent meeting. The new director was quite frank, saying that Conservation had mishandled this file from the beginning. He suggested they would do an environmental assessment of the colony's MMP. What a breath of fresh air.

On December 22, 2010, the cover was still off the manure storage facility and Conservation admitted they didn't know if the alterations the colony was making on the first compartment of that facility had ever been finished. They would find out and let us know.

After that, things got markedly better in our dealings with Conservation. We met with an officer from the Ombudsman's office regarding our complaint with the RM of Muddy Woods, and were happy with the outcome of that meeting. It seemed that she really wanted to help us.

***

ON JUNE 14, 2011, I NOTICED A lot of piping lying on the ground just south of the hog barn. I contacted the RM and was given information on a new project starting at the colony. I discovered they were in partnership with Manitoba Hydro to build a biomass heating facility, which included a research project. I contacted Hydro to get more information, which led to a stormy meeting with the two representatives from Hydro. I made clear to them our lack of confidence in the RM and the colony and their disregard for government regulations. One attempted to be sarcastic with me, and that was a mistake. I don't think he enjoyed his evening that much and had nothing more to say.

I also contacted Conservation and discovered that neither the colony nor Hydro had acquired the necessary permits for this development. You would have thought that Hydro, being a crown corporation, would automatically look after things like that. However, in discussions with a former Hydro employee, I was led to believe that this was a common occurrence with them. We talked to our MLA about the situation, and he made some calls.

Before long, a stop was put to that construction. Those pipes lying on the ground had alerted us to yet another unpermitted project going on at the colony. The project was held up for a year while Hydro and the colony obtained the necessary permits. I must say, I felt pretty good about that.

There was a provincial election in the fall of 2011, and on September 12, we met with our MLA. He was well aware of the problem and had been involved from the beginning. He was in opposition and so could more easily criticize government actions and policies. At the meeting, we tried to pin him down to exactly what he would do for us if the Conservatives were elected. He was an experienced politician and wouldn't commit himself to resolve any of the major issues. That's when we gave him an earful about what we thought of government regulations that protected the producer and gave virtually no rights to affected members of the community.

Next we met with the NDP candidate. It was an interesting meeting, and she seemed to understand our concerns. She agreed with us that the government's regulations were not good for the community or the environment. However, because we were in a 'yellow dog constituency'—even a yellow dog would be elected if it was running for the Conservatives—where that party had been elected for the past thirty years, she was pretty safe because she knew she was unlikely to be elected.

I complained to her that I could not get the ear of the various ministers involved in Conservation or Agriculture. She suggested to me that if I was to join the NDP, I would be much more likely to get an audience with them. So I joined the party. When I did that, I told my group I had taken one for the team. Besides myself, I think there might be still two or three other people in our group who support the NDP party of Manitoba.

***

ON JULY 15, 2011, I REPORTED ODOUR coming from the manure storage facility to the FPPB. The new inspector visited the site twelve days later. The colony was told to pump off the surface and consult an engineer. We also complained to Conservation about the lack of progress on getting a proper cover on the facility, and were told the colony was working with an engineering consultant to address the issue.

Shortly after that, we got a nasty letter from the RM of Muddy Woods informing us that there were no additional bylaws, resolutions, or permits that reflected the change from 658 AUs to 889. In other words, it proved that the RM had not passed anything that allowed the colony to make that change. The Ombudsman had clearly told them to give us that information—it was not something the RM would have willingly given us because they knew we could use that information against them. At least we were successful in getting *that* from the RM.

I contacted the Director of Environmental Operations of Manitoba Conservation and asked what the status of the environmental assessment was at the colony. He indicated that the colony would be assessed the following year, and that they had not yet sent in the engineer's environmental assessment for their human waste lagoon. Conservation told us the colony would not comply until they were threatened with legal action. They clearly did not see the laws of the land as all that important.

On February 10, 2012, the colony hired a consultant for the human waste lagoon—six-and-a-half years after the problem had been discovered.

***

THAT FALL, OLD JOE AND I PULLED the motor out of the 1950 Ford one-ton pickup truck. We took the engine apart and took the block into a machine shop to have it checked over. It turned out that the block was cracked. Fortunately, Joe had another one back in the bush. We hauled it out, took it to the machine shop, and had them rebore it. In went the new pistons and rings, bearings and new valves. It was like a new engine.

Joe then decided that he wanted to do something with the other truck we had overhauled. For a while, he was talking to a recycling company in Vancouver who wanted it for their business so they could go around and pick up recyclables. Joe was intrigued with that idea, but they weren't able to come to an agreement. It was going to cost more than a thousand dollars just to have the truck shipped out to B.C. So Joe approached the Threshermen's Museum in Austin that I had first visited when I was fourteen. They were delighted to get it and gave Joe a very healthy tax receipt. That truck is on display there every summer, usually parked beside the grain elevator on the grounds. I go and say hello—to both Old Joe *and* the truck—every time I attend the Threshermen's reunion. I am pretty happy about the old truck on display as well as Old Joe's lasting gift of friendship.

On January 10, 2012, we met with the two neighbouring RMs—Greenland and Snow Valley—seeking their support and involvement in asking for a count of the pigs at the colony. Most of the people affected by the odour, including myself, lived in those two RMs. Both were quite supportive and approached the RM of Muddy Woods about conducting a count. With such a show from the community, Muddy Woods had no choice but to agree to count the pigs in the colony operation. That count took place on November 12, and as far as we know, it was both the last time they did a count, and the last time an independent person checked for compliance.

After the count, Muddy Woods met in camera with the other two RMs. For one reason or another, they did not want us to know what the actual count was. At some point, they had been told by their lawyer not to talk to us because it was just a matter of time before we would sue them.

We found out what the count was anyway—well over 1,200 AUs. Were we surprised? Not by any means, believe me.

CHAPTER SIXTEEN

Bringing in the Sheaves

IN THE FALL OF 1960, DAD GOT an infection in his salivary glands and decided to go to the local hospital. They gave him a shot of penicillin and sent him home. Within a few days, he got a rash over most of his body and his face swelled up. He went back to the hospital, and they admitted him; he had an allergic reaction to the penicillin.

This happened just at the beginning of the harvest, and our family was threshing with the Carters and Smiths. Mr. Carter owned the McCormick threshing machine and drove it with a styled John Deere AR. It was a good outfit, and Mr. Carter ran it well. We still had some sheaves lying down, and I stooked the rest of the crop that we had cut with the binder.

School had just started, and I went to the teacher and told her what had happened to Dad. I said that because of that, I had to do my family's share of the work in the threshing gang. Normally, we didn't start to thresh until after lunch time anyway. The grain was still damp, and by noon it had dried enough so that it wasn't tough. I told her that I would like to leave school at noon to thresh at the Carter's. At that time, there was a law in Manitoba that allowed farm boys two weeks off from school to help with the harvest. I was her best student, and she was not concerned about my grades, so she agreed. The school was only half a mile from the field where we were threshing.

That day after school let out, I went to help at the Carter's as usual. I approached Mr. Carter who was greasing the machine, and told him I would come at noon the next day and bring our team and rack. He turned to me and said, "I don't want you to do that, Billy. You are a good student, and you should stay in school." It was a small community, and everybody knew I was the top student there.

"But, Mr. Carter," I said, "my dad is sick, and we'll be threshing at our place next. I need to do our family's share of the work in the threshing gang."

He became a little annoyed, and told me, "The work will get done, lad. Listen to what I'm telling you."

I knew the discussion was over when he turned away to finish what he was doing.

The next day, I drove the team and rack to school, unhitched the horses, and put them up in the little barn that was on the school yard. I waited till the last recess around 2:30 in the afternoon, and looked at the teacher who nodded to me. That was all I needed. I went out with the other children, and when they went to play ball, I got the horses out, hitched them up, and was on my way. I trotted the team over to the far side of the field at the Carter's where we were threshing. It was as far as I could get from Mr. Carter. I was afraid that he would send me back to school.

Because I was gathering sheaves, I had the reins of the horses attached to the front of the rack. When you drove horses that way, you would use your pitch fork to press down on the reins to turn them this way or that. They would stop or go on voice command. It wasn't working that well until Mr. Smith came over to my side of the field and told me the reins were too tight. I loosened them up, and after that, things worked fine. I had timed it so when I brought my first load in, it was just after four. School ended at four. I waited for the team ahead of me to clear the threshing machine, and then I pulled up to unload.

I braced myself when Mr. Carter turned towards me; I wasn't sure what he would say. He looked at me and muttered, "I guess I've nothing to say!"

I pitched off and trotted the team back out to the field to load again. I hauled another two loads before it became too dark to continue. Four loads was a good day for a *grown* man.

Mine was the last load we threshed that evening. We shut down the outfit and went into the house to have supper. Mrs. Carter's kitchen was warm and beautifully painted in white with red trim. Delicious smells were coming from the old wood cook stove. Mr. Carter and Mr. Smith invited me to come and sit with them at the table. Mr. Carter said to me, "You did the work of a man today, Billy." I had come of age. In my heart, I knew I was okay, and that wherever I went, I would be at home and accepted.

For the next two weeks, as long as Dad was in the hospital, I would leave school at noon and take my place in the threshing gang at the Carter's, at the Smiths, and at our place. I was still doing chores before school and after I got home from threshing. I was tired, but content.

***

DAD KEPT HAVING RELAPSES OF HIS ALLERGY to penicillin. There didn't seem to be much they could do for him, so he just had to sit in the house and do nothing. This was not pleasant for anybody. It didn't affect me too much because I was outside most of the time doing chores, cutting firewood, and riding the horses over to my friends. But you could feel the tension rise in the house every time a relapse happened. Mother complained even more loudly and longer than before, and a concerned and worried look was on my sister's face most of the time. Dad was like a silent smoking volcano ready to erupt at any given moment.

It was the beginning of another tough winter. We were living off pork and potatoes again with a bit of mutton from time to time. I had cut some wood that fall, but we were out of wood by Christmas. A fire had gone through the Moylan's place, and there was some lovely fire-killed poplar that was the right size and easy to get. When he was well, Dad and I would hitch up the team and sleigh on any bright, sunny winter Saturday, go over there, and cut wood. We would take some pork and homemade bread, if there was any, and build a fire and toast our food at noon. The snow was at least a foot deep, and I enjoyed watching the fire sink down to the ground. It looked like summer down there and you could see the leaves and grasses. Dad would start telling me about his experiences during the war as we sat around the fire eating lunch.

Dad had been a corporal and a scout for the Fourth Armoured Division. He had spent his youth hunting deer and had a wonderful sense of direction; I can honestly say I have never seen him turned around and become lost. When the army discovered his skill, he was given a Harley Davidson so he could guide convoys of trucks and tanks over mine-seeded roads in France and Holland at night. Apparently, he was pretty good both in guiding the convoys and avoiding the mines. He told me about how they would repair their trucks on the road by scavenging parts from abandoned vehicles. He even told me about a few of the incidents where he had seen men killed. The men he talked about were people he didn't know.

In November 1944, he got sick with infectious hepatitis and was airlifted back to England in the Halifax bomber. These bombers would land in Holland after attacking Germany and pick up the sick and wounded. Dad had aunts and uncles who still lived in England, and he visited them while he was there. When he described what England was like in winter, he said it was the coldest place he'd ever been.

Dad was part of the occupation force in Germany after the war and didn't get back to Canada until 1946. Apparently, the Allies did not have enough ships to bring the soldiers back before then. The soldiers who had gone over there first, like my uncle, were the first to get back. The ones who'd gone over during the war waited until ships were available. Dad never spoke about what happened when he got back until years later.

When he was done telling stories, we would finish filling the sleigh with wood before heading back to the house. We were always on the lookout for a wolf because that meant money for groceries, so we always brought a rifle with us.

One day, he brought the old .303 Enfield he had got from Mother's father. It was a First World War relic that had been shot out years ago, and not the most reliable. When two gray wolves showed up about two hundred yards away, I asked Dad to let me hold the horses while he shot at them. Sometimes, he was just senselessly stubborn, so instead of handing me the reins, he tried to hold the horses and shoot at the same time. When the first shot rang out, the horses reared and pitched, and he couldn't get off another shot. It probably didn't matter much because with that old rifle, the bullet could have gone anywhere. He did much better with the .22, but of course the wolf had to be closer. After that, we took the .22 to the bush with us.

When we got home, we would pull the sleigh-load of logs into the yard, and I would bucksaw wood every day after school. When that was done, I'd cut another wood box full.

I was anxious to hunt as well, but because I was so nearsighted, I couldn't aim a rifle with open sights. There was no such thing as telescopic sights in those days. My parents knew I needed glasses, but didn't seem interested in getting them for me. Dad was still smoking at the time and had gone to roll-your-owns to save money, but he never seemed to run out of cigarettes. The subject of my glasses was never brought up, and for me, it just seemed to be a way of life, like having no money. I didn't think about it much.

My sister was spending more and more time at the Pentecostal Church in Glenella. I was getting worried about her because I could see that this was no answer for her problems; she was now talking that fundamentalist rhetoric. She wanted me to go with her in the worst of ways, and it became an issue between us. Looking back on it now, I realize that it was probably better for her than nothing.

Aunt Evelyn arrived that fall with a care package of hand-me-down clothes, baking, and early Christmas presents. Father hardly spoke to her while she was there. I heard later that my uncle had asked her if he was supporting Bill Massey's family as well as his own. I don't remember her ever coming again. We went to the neighbours' that year for Christmas. At least we didn't have to take that long and lonely drive through the wind and snow.

CHAPTER SEVENTEEN

Losses and Gains

IN NOVEMBER 2012, CONSERVATION TOLD US THAT the colony had not met their requirements for the human waste lagoon. Conservation was working on a prosecution package, after which the colony decided to comply. In their culture, they kept themselves apart from the community, and it seemed that they did not see the laws of the land as important. The fact that they were having an effect on the community around them did not seem important either. As a result, none of the members of our group or other community members were having much to do with them anymore. The colony was isolating itself.

When 2013 arrived, it was the ninth year of our struggle with the colony and the rural municipality (RM) of Muddy Woods. On February 25, we heard from Conservation about the manure storage facility (MSF). The MSF had been constructed in 2001. In 2006, the colony had made unengineered modifications to it and was not permitted by Conservation to put the cover on. We learned about this situation at our meeting with the colony on April 29. The heavy cover, a sheet of material as large as the surface of the MSF, went back on in 2008 and was soon covered with effluent. In 2009, we had raised concerns with the Farm Practices Protection Board (FPPB) about the odour from the MSF. In 2010, the FPPB told us they had no mandate to deal with the odour because Conservation held permit on it. That meant

Conservation had temporarily taken over management of the facility and was overseeing the situation until the engineering assessment was done and the modifications were approved.

There had been a dispute between Conservation and the Manitoba Farm Industry Board (MFIB)—which had replaced the FPPB about that time—as to who was responsible for the odours from that structure. This disagreement had gone on for years while these government departments argued over who would enforce the requirement to cover the MSF. It was finally determined to be the MFIB in February 2013. Imagine two government agencies squabbling for three years over who was responsible for this situation, while we sat in the stench.

In March 2013, compliance in the hog barn had been achieved, as evidenced by a count completed on Muddy Wood's behalf in mid-March. According to our sources of information, the count was reported to be 979 animal units (AU), which is exactly 889 plus ten percent. Amazing that they could hit that number exactly. I don't know who did the count for the RM, but I have my suspicions. We asked the RM for the numbers for both counts in the fall of 2012 and in March 2013, but of course they refused, so we launched a complaint to the Ombudsman. Unfortunately, the Ombudsman would not support our request on the basis of Section 18(1)(b) of the regulations of the Freedom of Information and Protection of Privacy Act. Essentially it said we were not allowed to see that information because we might use it to compete with the colony in the hog business. As if.

We made a written submission to the Approvals Branch of Manitoba Conservation and Water Stewardship about our concerns regarding the new sewage lagoon. We had suggested that the colony hook up to the new municipal lagoon system that was being constructed for Big Island; we were concerned about the number of lagoons and MSFs in the area and the effect they might have on public health. Another issue was the migratory wildfowl that travelled through the area. The concern was for the birds themselves, as well as the transportation of harmful bacteria. We also talked about the apparent lack of concern that the colony had demonstrated for the surrounding community. We had no confidence that the colony would follow the regulations in their management of this new sewage lagoon, and the mishandling of these situations by Conservation over the past number of years had done nothing to reassure us that our interests were being looked after.

We were in close contact with the Conservation officer responsible for this development, and met with him several times to discuss our concerns. He was a local person and very familiar with our group.

***

TOUGH OLD JOE WAS HAVING SOME HEALTH issues and was eventually diagnosed with cancer. He had sold his property to his daughter and moved to a house in Stonewall. He still remained active with the Concerned Citizens of Big Island (CCBI), and frequently came out to Big Island to help his son. I was working for Young Joe from time to time, and the three of us often worked together. Old Joe was still going to Texas for a few months in winter, and I usually helped him get his motor home ready to go. That summer, he was admitted to the hospital, and I went to see him several times.

One day as I was driving by, I thought I should stop in and see him, but I had my dirty work clothes on and decided against it. I will always regret that. It was one time I didn't listen to my instincts.

Early the next morning, the phone rang, and Dorothy answered it. It was Young Joe, and he asked for me. He told me that his dad had died that morning, and he was heading to the hospital to take care of things. I put the phone down, and felt a tremendous loss. How does that saying go? When a giant tree falls in the forest, it leaves a huge gap. And that described exactly how I felt.

Old Joe had also had a pretty rough childhood. He was more than a survivor; he was a warrior, like me. We had become fast friends, and I relied on him a great deal in this struggle we were having with the hog barn. He was a main source of strength for me. It was one of the few times in my life that I shed tears for the loss of someone.

Young Joe stepped forward to fill the place that his father left behind on CCBI. I was so appreciative of that.

***

NOT MUCH WAS HAPPENING ON THE HOG barn front, but by September 2014, the smell from the barn was again atrocious. I contacted the MFIB, and Young Joe sent a supporting letter describing the smell in his yard. In their response, the MFIB gave me the contact information for the RM of Muddy Woods. I thought that was kind of funny.

On October 15, CCBI met and decided to request an immediate count of the animals in the hog operation at the colony. The next day, one of our group called and told me she had seen uncomposted animal remains spread in a field just west of the colony building site. I went over there and took pictures of not only chicken, turkey, and pig remains, but also a third of a carcass of a cattle beast.

I reported the situation and sent pictures to the local Conservation district office, and also wrote a letter to the MFIB. The latter indicated they would send a livestock specialist to the colony to investigate our concern that the order written in 2007 had been violated. The colony has their own composting bunker that they empty in the fall. Stray dogs and coyotes have been known to feed there and the young people trap and snare them. They have also set traps and snares on private property, in some cases without permission, and deer sometimes get caught. In one case someone's companion dog was killed. Conservation investigated, but could not determine who the culprit was.

Conservation attended the situation at the colony on October 18 and told them to clean up the remains immediately. Within a day, the entire colony—women, children, and even some of the men—were out in that field picking up uncomposted animal remains. I felt sorry for them, but perhaps it was an event that was needed to change the colony's behaviour. I hoped they were taking precautions, such as rubber gloves, so that no one would get sick as a result.

On October 24, the colony was charged by Conservation with the improper disposal of uncomposted animal remains, and fined $2,470. Finally, after all these years, some action was taken.

On November 19, we received a letter from the RM of Muddy Woods respectfully declining our request to perform another count of the hogs at the colony, the extreme smell notwithstanding. The letter said they did not believe that the colony was in any violation of the existing municipal bylaw. They also stated that odour and farm management practices were not within the purview of the RM.

It seemed strange that in our search for some government body to help us with our problem with odour from the barn, nobody seemed to think it was their responsibility. Not the MFIB, not Conservation, and apparently not the RM. They had also indicated that they would not consider any more correspondence from us in relation to this matter. That did not appear to me to be very respectful. I'll spare you the disrespectful response that came to *my* mind.

***

BACK IN FEBRUARY 2014, I BECAME PRESIDENT of the NDP Constituency Association in our electoral division of Lakeside after two years being a member of the party. In that capacity, I attended a meeting of the Agricultural Committee and happened to have lunch with the Minister of Agriculture. I took this opportunity to describe to him in detail the situation as it had developed over the past ten years. I don't think that he particularly enjoyed his lunch, but he listened carefully to what I had to say.

Outside of sympathy, I don't think I gained much from that conversation. However, I had joined the party and became the president in order to *get* these opportunities, and I knew that more would present themselves in the future. I talked to the Minister's office after that meeting and asked for a review of the situation. I got a very nice reply, stating that my concerns had been brought to the Minister's attention, and he thanked us for sharing the information with him. There was no mention that he was prepared to conduct any sort of review.

The Minister's office did, however, arrange for the livestock specialist investigating the poultry manure issue to look into the situation with the hog barn.

Municipal elections in the fall of 2014 resulted in some changes in the council. On December 10, our group met and decided we would talk to the new council.

***

EARLY IN JANUARY 2015, I WAS CONTACTED by the Conservation officer, Bruce Webb, who was in charge of overseeing the colony's proposed facility that was to replace their human waste lagoon. He offered to meet with us to discuss the situation. That led to several other meetings with him throughout the process, from approval of the project and the start of construction in the summer, right through to completion. That development cost the colony close to a million dollars. Everyone in CCBI was very pleased we had at least accomplished this for our community.

We had been requesting the colony's Manure Management Plans (MMP) twice yearly from the beginning. I can't remember what caused me to ask the Conservation officer in that department if there were any penalties for misrepresenting information on them, but I remember how surprised I was to hear there were no penalties. It was then I realized that the colony could be putting any numbers down on the MMPs to hide the number of pigs that were actually on the colony. Not only that, but it could mean that the part of the MMP that involved the spreading of manure could also be fabricated. If that was the case, then there could be a real problem with levels of nitrogen and phosphorus on the spreading fields. I stopped asking for those plans at that point because I felt the information they contained might not be valid anyway. I also wondered if that was the colony's plan all along; constantly wading through all the misinformation provided so far was certainly wearing us down.

We asked Conservation to compare the number of AUs between the 2012 MMP and the official count of AUs that took place in 2012. Sure enough, they did not coincide. Conservation said that with respect to The Planning Act, it still remained in the RM's hands to deal with any discrepancies in the number of AUs that are housed in a facility. This was the responsibility the RM had abdicated.

I remember the colony complaining about the bureaucracy and all the forms they had to fill in for the government. In a way, I didn't blame them because it was apparent that nobody was interested in checking any of the information on those forms. And if a serious problem did develop, it was in the hands of the RM anyway, and they didn't seem willing to act on it at all.

It seemed like the whole thing was a façade to make the government look as if they really were keeping a handle on the situation with these barns. I could understand the colony's frustration because they knew there were no consequences for misdemeanours anyway, and they were filling in mountains of paperwork to maintain the image the province was promoting about their regulation of the hog industry. It really was true that this province had some of the toughest regulations, but they were seldom enforced, and we're still working to change that.

CHAPTER EIGHTEEN

I Could See, See at Last!

THE WINTER OF 1961/62, I WAS OUT riding the horses quite a bit. My friends at school were two boys in my grade, Allen and Reggie, and Andrew Cole who was younger than me. Andrew was in a Catholic school and didn't attend ours until Grade 4. I'd ride over to my friends' places and pull them in the fields on a toboggan behind the horse. By this time, I was riding the mare; she was as fast as the wind and loved to run. I also really wanted to hunt, and I found if I squinted hard, I could see well enough to shoot a bit with a .22 at close range.

I especially liked my teacher, and she kept bringing me books on ancient history. By Christmas time she was engaged to be married. I wasn't sure how I felt about that, but I continued to be a good student and worked hard at school.

There was a blow dirt bank that ran along the south headland of our quarter. During the drought in the '30s, the wind had picked up the worked soil and blown it onto road allowances and fence lines. It created a drift of soil similar to a snowbank. The quarter a mile south of us had been ploughed one spring back in the thirties, and the wind took the soil down to the depth that it had been ploughed. That quarter was ruined for grain. The Taylors owned it, and we made hay there instead.

I remember sitting in our house one day and suddenly seeing a timber wolf going east on Mr. Cole's field just to the south of us. It was sometime in February, and that winter the timber wolves were coming down from the Riding Mountain National Park to find food. They must have taken most of what was available in the park.

When I saw the wolf, I quickly jumped into my boots and grabbed the .22 rifle. By the time I got outside, the wolf was on the other side of the blow dirt bank; I couldn't see him, but I knew he was there. It was stormy, and the wind was blowing from him towards me, so he couldn't hear or smell me. However, I knew I had to get to the end of that bank and be ready to shoot before he figured out I was there. There must have been about eight inches of snow on the ground, which made it tough to run through. The wind took my breath away, and the snow stung like grains of sand hitting my face. My lungs felt they were on fire by the time I had run that half mile. I dropped to one knee and raised the rifle just as the wolf came out from behind the blow dirt bank. He was close enough, and I squinted hard to make a good shot. He dropped into the snow. I quickly put another round in that single shot rifle in case he got up, but he didn't.

Needless to say I was pretty excited and proud of my accomplishment. He was so big I couldn't lift him, so I walked back home and got old Trigger. I didn't want to bring the mare because I thought the sight of the wolf would spook her. I put a harness on the horse, and when I got there, I tied a rope around the wolf's neck and attached it to the traces. I led Trigger home, dragging the wolf, and stretched him up with a wire stretcher to an oak tree. Hoisted up like that he was at least six feet long. When Dad got home, we skinned him out. We did pretty well with that hide, which kept us in groceries for several months.

***

THE OLD BARN WAS STARTING TO FALL down and was so full of manure it wasn't usable anymore, so that winter, we built another barn out of material salvaged from the old house. It had a flat roof, so we put a foot of hay up there. There were bales piled up against the walls and it was quite warm inside. Dad had a headlight bulb he'd taken out of the

'40 Chevy and that was our light. We kept the battery in the house to keep it warm and carried it to the barn in the evening when we were doing chores.

There's nothing like a barn in winter time at night. The animals greet you with pleasure, and the smells are wonderful. It's just the best place to be. I loved going to the barn that winter.

We had an ancient Billy goat. Dad had gone to a sale with the team and rack, and when the sale was over, he saw that somebody had tied the goat to the back of his rack. At least that was the story, but I really think Dad agreed to take it because it was clear no one else wanted that old goat, and that's how it came to live with us.

That goat proved to be the bane of Dad's existence. It had one long horn that curved right around and almost touched the ground. The other horn had been cut off about halfway. The goat would jump into the pigpen and use his horns to butt the pigs out of their trough and eat the grain. One day when it did this, Dad was waiting. He roared at the goat, and as it fled out the barn door, Dad swung the fork in rage and frustration and connected with the one good horn. After that, with only half a horn, the goat was not as efficient in getting the pigs out of the trough.

I've never smelled anything as bad as that goat. I think the huge growth hanging from its neck was the source of most of the smell. I have no idea how old the goat was when it came to our farm, but I'm sure it was well aged. Someone brought a doe over to be bred, but the old goat was just not interested. It lived with us for several years before finally dying of natural causes.

I had started taking the '40 apart because I thought we could make a rubber-tired wagon of the wheels and frame. All we had were steel trucks, and I liked the rubber-tired hay wagons that the Taylors had. I didn't get too far with that project because I decided to use the rear end of the car to make a tractor. Dad had picked up an old Massey Harris hit-and-miss engine at a sale—one and a half horsepower for five dollars. It had spent most of its life running an air compressor in an elevator. Mr. Cole had some pieces from a Model T Ford in his bush, and told me to take what I wanted. I got the front axle as well as a coil that was missing from that engine.

I remember taking apart the differential in the rear end of the '40 because I wanted to know how it worked. It was pretty easy to understand once I got a look at the spider gears. I worked on the hit-and-miss engine, putting a pulley on the end of the drive shaft, and using a belt with a tightener for a clutch. I also built a platform on the back so I could haul a barrel of water up to the pasture for the horses and cattle.

I was usually the first one out the door when school was over because I had things to do at home. I had noticed as I was leaving school one day that Andrew was being pushed around by my two friends from my grade. He was three years younger than they were, and he didn't back down from anyone; he was a feisty kid. I could have stopped and gone back, but I really wanted to go home and get to work. The bullying had been going on since Andrew had come to our school, but it had recently escalated, and I admit I didn't want to get involved.

I was in the middle of hauling water to the pasture with my tractor when Mr. Cole caught up to me. From the look on his face, I suspected that Andrew might have gone home and told his dad about the bullying. Anyway, when Mr. Cole asked if I had seen anyone hitting Andrew, I said that I hadn't. I explained that was because I was always the first out of the school, and it probably happened after I left. I convinced myself it wasn't exactly a lie, just a play on words, but I still feel a twinge of guilt when I think about it now. That day I learned the importance of asking the right questions and telling the truth.

When Mr. Cole asked if I would keep an eye on Andrew at school in the future, I said that I would. Mr. Cole was a big angry man that day, and I was glad he wasn't looking for me. The next day, my friends told me that Mr. Cole had come to their homes and talked to their parents. I gathered from what they said that it wasn't very pretty, and their parents had told them afterwards to leave Andrew alone. I told them that Mr. Cole had asked me to look out for Andrew, and agreed that they had better leave him alone. They both knew me and knew that I had taken care of a bully who was bigger than me. As far as I know, they let Andrew be after that.

It was another busy summer making hay at the Carter's, at the Taylor's, and at home. I was also helping Mr. Cole that summer with haying. He had a nice Massey Ferguson baler that dropped square bales on the ground. Andrew drove the H John Deere, and Mr. Cole picked up the bales with a fork and put them on the wagon. I remember he had trouble because he couldn't find a fork handle strong enough to not break while he forked those heavy bales. Mr. Cole was very strong, but he was bothered with sickness, so he bought a little elevator and fastened it to the side of the rack to make picking up the bales easier. Andrew and I could easily use that outfit to do all the work ourselves, and I helped haul a lot of bales for Mr. Cole that summer.

I remember going with them to Austin, Manitoba, for the Threshermen's Reunion. We all went in their 1950 Fargo one-ton pickup. Packed into that cab were Mr. and Mrs. Cole, Andrew, their two little girls, and me. If you've ever ridden in one of those old Fargo one-tons, you'll know it's not a smooth ride, but those parades were well worth any discomfort. I've been there many times since and still see some of the same tractors going by today. Mr. Cole bought us lunch and we had a great day. Mr. and Mrs. Cole were really nice to me.

***

I HAD EARNED ENOUGH MONEY THAT SUMMER that I could afford to buy myself glasses. I remember the day I got them and put them on. My world changed instantly because I could see. Everything was so clear and distinct. The first thing I did was grab the .22 and hunt grouse along the wooded road allowance to the west of us. It was a beautiful fall afternoon and the leaves had changed colour. Hunting was good, and I got several birds. Things were so much better because I could finally see.

My mother was pregnant that summer and expecting her baby in November. I hadn't said much to her since I overheard her comment about Marlene and me a year or so before; I don't think she really noticed the lack of conversation between us anyway. One school day in late October, when Dad was away and Mother was home alone, I decided I should go home and check on her, so I asked the teacher's permission at lunchtime if I could run home. When I got there, Mother asked sharply, "What are you doing here?" I told her I'd come to check if she was okay. She said somewhat casually, "That wasn't necessary. You should be at school."

It was over a mile to the school, and because I was a little late getting back to class, I went up to the teacher and apologized. She said, "That's alright, Billy," and I noticed she had tears in her eyes. The look she gave me and the tone she used made me feel like a million dollars. I didn't need my mother's approval when I had hers.

Later on, I found out that she had just learned that she was also pregnant. I suppose she was moved by what I had done.

CHAPTER NINETEEN

Honour and Deceit

IN FEBRUARY 2015, I SAT DOWN WITH Stuart, my councillor from the rural municipality (RM) of Snow Valley, to discuss the hog barn situation. He came over to our farm, and after we spent some time talking, I drove him around the section where the hog barn was located, and he asked me to call him when the odour was quite bad at my place and in Big Island. Indeed, when the winds were blowing in the right direction, he came by to experience the smell for himself. His brother, who happened to be the chair of the Pork Council, had also experienced the smell in Big Island.

Our group, the Concerned Citizens of Big Island (CCBI), then decided to ask the CAO of Muddy Woods for an information session with the new council and reeve. My councillor, Stuart, agreed to meet with the incumbents who were still on the Muddy Woods Council. A councillor who represented the people from the RM of Greenland teamed up with Stuart and went to visit Peter at the colony. Our purpose was to find common ground with the RM and the colony in order to reach a solution to maintain appropriate numbers in the hog barn. These councillors took their responsibilities seriously and sincerely wanted to help their constituents. We were encouraged and grateful for their support and involvement, and I was happy to let Stuart take the lead in this effort.

In early March, our group did another presentation to the RM of Muddy Woods. I went through a list of events and actions from the beginning of June 2004 till the end of February 2015. I had made the list as brief and concise as possible, but it still contained forty-nine items. One of the incumbents, who did not support us anyway, fell asleep during my presentation. That really didn't matter, as we were pretty sure the RM didn't want to do anything at the time.

I concluded my presentation by suggesting this was a very complicated issue. I said if they required further clarification and discussion, I'd be happy to meet with the council again or with individual councillors to discuss any questions and provide more information. We urged the council to form a committee with ourselves and the other two RMs so we could all sit down and come up with a solution to this problem. A month later, we received a polite thank you for the presentation, and a couple of weeks after that, another letter declining to be part of the proposed committee.

The two councillors from Snow Valley and Greenland decided to try to go it alone, and met with the colony several times that summer. The colony actually seemed receptive, and the two councillors hammered out an agreement with Peter and Sam, which they presented to us late that summer. CCBI was not all that happy with the agreement because we didn't feel it went far enough. We then discussed it amongst ourselves and decided that at least it was better than what we had at the present, which was nothing. Those two RMs drafted the agreement and sent it off to the colony on September 14, 2015.

The terms of the agreement were as follows: (i) A three-day notice would be given for a count; (ii) Personnel recognized by Manitoba Pork would complete the count; (iii) Should the colony be in compliance, CCBI would cover the costs. Because we really wanted to get some kind of agreement, we were prepared to do this, but Stuart himself offered to pay for a count; (iv) Another count would not take place for two years. We felt this was the weak part of the agreement because the colony could increase the herd within six months; (v) If noncompliance occurred, the cost of that count would be borne by the colony, and a subsequent count would take place within six weeks to confirm compliance; (vi) Another count would then take place within one year; and (vii) A signature on behalf of the colony was required by December 31.

We all waited for the colony's response.

***

BACK IN 2014, I HAD JOINED THE NDP Constituency Association of the Interlake electoral division, which was just to the north of us. Several members of the Association attended the meetings on a monthly basis. It was very interesting, and I learned a lot about what was going on in the party and the government in Manitoba. The MLA, Tom Nevakshonoff, was interested in our situation and open to helping where he could.

After I became President of the Constituency Association of the Lakeside electoral division, a member of the NDP contacted me about a situation near to where he lived in the Northern Interlake. There, fields had been created out of bush on very poor, thin soil. Two hog barns had been built across the road from each other, and the manure was being broadcast in the fields. About half a mile south of the hog barn was an artesian well. When I saw it, the water was gushing out of the ground, flowing at the rate of a broken fire hydrant. The water flowed into a municipal ditch that ran between the two hog barns and the spreading fields. Surface water from the fields drained into the ditch. After several miles, the water ran into the Icelandic River, and several miles further, into Lake Winnipeg. When I talked to Tom about the situation, he was under the impression that well had been capped. Apparently, an attempt had been made, but was unsuccessful. Once I reported it to the MLA, it was out of my hands. I have not gone back since so I can't say what happened.

In 2015 the party had got itself into trouble, and there was a lot of dissension in the ranks and unhappiness with the leader in the cabinet. There had even been an unsuccessful attempt to replace the premier, and recent polls showed that the party was losing ground. We were not surprised when the NDP lost the next election.

That spring, before the election, Tom was made Conservation Minister. The provincial environmental assessment of the hog operation at the colony promised by Conservation in 2012 had never happened, but in 2015, Tom made sure the assessment took place. From what I saw of it, I don't think anything was discovered to be wrong with the colony's operation. And although there had been no count of the pigs made at that time, it was at least something that I had achieved for our group.

***

ON OCTOBER 9, WE FINALLY RECEIVED THE report from the MFIB investigation that had taken place back in October 2014 following the incident where uncomposted remains were spread in a field adjoining the colony building site. The MFIB had agreed to check the hog barn, as well as the incident with the spreading of uncomposted animal remains, but by the time their investigation had taken place, many of the remains had been removed.

The report suggested that the requirement for the colony to inform the community about any upcoming land applications of manure might not be desirable considering the long-term conflict between the community and the colony. We were astonished by that statement. In other words, the colony was given permission not to communicate with the community. It was interesting how the problem had become our complaints rather than the smell or the colony's behaviour.

The one thing the report did yield that was important to us was a revelation that the colony had 900 sows, farrow-to-finish, for a total of 1,125 animal units. We discussed this and somewhat reluctantly decided that we would wait until the agreement with the colony was signed before we would bring this to the attention of Muddy Woods. We did not want to do anything that might jeopardize the signature to that agreement, weak as it was.

December 31 came and went without any word from the colony. The two councillors went to the colony and urged Peter and Sam to sign the agreement, but they simply responded that they were still considering the agreement carefully.

On March 7, 2016, my RM of Snow Valley wrote to the colony indicating they had not received a response in regards to the proposal they had presented. They assured the colony that their concerns and those of the citizens of Big Island that they represented were not going away. I liked that language.

On April 20, Snow Valley gave the colony a date by which they expected their response, and said that if none was forthcoming, they would present our concerns to the Pork Council. My councillor, Stuart, who I had come to regard as an upstanding and highly respected man with a strong moral compass, went back to the colony and made a personal appeal to Peter. During the course of that conversation, he asked Peter how he was able to sleep at night. Peter offered no response. Apparently, it was not at all a concern for Peter, which made it clear to us that this had been a stalling tactic all along to keep us off the colony's back for almost a year.

And then we heard that back on March 22, Muddy Woods had a Conditional Use Hearing where they approved a proposed feed mill development at the colony. They did not inform CCBI of the hearing, nor did they inform the members of CCBI who were entitled to that information. When we later confronted them, they insisted they had.

This story was far from over.

CHAPTER TWENTY

Frostbite and Friendship

NOVEMBER 22, 1961, WE WERE PLAYING FOOTBALL at recess when suddenly our old '47 came flying down our road, turned down the provincial road in front of the school, and went roaring by. I thought that was rather unusual behaviour for my father, although I had seen him abuse his car in the past when he got angry.

When I got home from school, neither he nor Mother was there, and Marlene and I figured out that the baby had come. I went about doing the chores, Marlene stoked up the fire in the house, and we made ourselves something to eat. Eventually, Father got back and told us the baby was born and it was a girl. I heard later that she had come very quickly, which was the reason he was driving that old car so fast. Apparently, Mother barely got inside the door of the hospital, and the doctor didn't even have time to take off his coat before he delivered the baby. I think that was the only time my little sister Jean was ever in a hurry.

I enjoyed having her around because in some ways, it was akin to having a puppy in the house. She liked me just like the puppies liked me, and she was usually happy to see me. The best part was she kept Mother busy, so Mother wasn't after me for things she wanted done. She was clearly happy being involved with the baby because she wasn't complaining as much as usual. I did my share of looking after Jean, changing her, feeding her, and putting her to sleep at night. She really didn't make much of a difference in our parents' attitudes towards Marlene and me, and the two of us went about our lives as usual. Marlene was spending a lot of time at her church, and I was over helping the neighbours.

That winter, I went hunting almost every day after school. We ate lots of grouse, but Mother refused to cook the jackrabbits I shot. She said she had eaten too many during the Depression and refused to cook another one. I shot several of them and lots of bush rabbits. I remember one day getting about a dozen, tying them all together with a piece of rope, and dragging them home. I threw them up on the roof of the porch, and every other day, I would bring one down, chop it in half with the axe, and give half to Shep and Queenie; they got the other half the next day. Those dogs never looked better than they did that winter, well fleshed out with shiny coats.

I was wearing a pair of moccasins that winter with holes in the toes. I had tried to patch them, but I wasn't successful and snow got in. One day it was bitterly cold and I went hunting. On the way home, I could feel my big toes freezing; it's not an experience I would recommend. At the point where they actually froze, I felt a sharp stab of pain. I didn't hurry after that as I knew there was no use. I even stopped at one point to do some target shots. I'm like that; when I injure myself, I simply continue whatever I've been doing. There is a side of me that is pure defiance, I guess.

When I got into the house, I went and lay down on my cot behind the organ. I was soon in agony as the toes thawed, and when I started to groan, Marlene came and tried to help me. When I sent her away, Mother said it served me right for going out on such a cold day.

Somehow or other, that story went around the district, and by the time it got back to me, I had apparently lost both big toes to frostbite. I do remember the blisters breaking at school one afternoon, and a little puddle of fluid leaking out on the floor. I quietly wiped it up without saying anything to anyone; I was embarrassed that it had happened at school. I had no feeling in the ends of my toes for almost thirty years. But finally the nerves regenerated, I guess, because they're perfectly normal now. The human body is amazing, isn't it?

I was laid up that winter while my toes healed. My teacher gave me a book about the First World War with pictures of warships. I remembered the stories that Grandfather told me and passed my spare time building cardboard models of battleships, cruisers, and destroyers. Some of them were pretty good; even my dad was impressed. I also listened to a radio program about Horatio Hornblower, a fictional captain in the British Navy during the time of the Napoleonic Wars.

The following summer, Mr. Cole had traded in his old John Deere D and bought a Massey Ferguson 88 gas tractor. I thought that was just the nicest tractor, and when he asked me to come over and help him, I was happy to do so. Mr. Cole was still bothered by sickness that winter and was looking after quite a few pigs, sheep, and cattle. Together, we belted up that '88 to the crusher and crushed a month's worth of chop for his livestock. I brought bales from the stack over to the barn and piled them up where it was convenient for Mr. Cole to feed the cattle.

I was over there a fair bit that spring until Mr. Cole and his doctor figured out what was wrong. Mr. Cole had an ulcer—he worried too much. Because their girls were still quite young, Mrs. Cole sometimes called me over to babysit so she could go with Mr. Cole to the doctor's appointments. I was happy to look after those girls because it also meant Andrew and I could play together with his Meccano set. When the parents got home, they always gave me some money. I know Mrs. Cole was grateful for my help with the family, and my friendship with Andrew.

That spring, our school had a pretty good baseball team. I was one of the heavy hitters, and being left-handed, managed several home runs. Competition was fierce amongst the various one-room schools in the district. Everyone in the community would come for our annual sports days, especially for the baseball games. In the final rounds, we were losing one very important game, and our team was getting pretty discouraged. By that time, however, I knew the value of positive reinforcement, and I started encouraging my team members.

We were in the final inning, and the other team was up to bat. I was the backstop, and in those days, we played without any protection at all. I took a foul tip off the bat, which hit me square in the nose. I was dazed and got to my feet, the blood pouring down my front. The umpire got me to lie down and applied pressure on my nose. People gathered around me, but my parents were nowhere to be found. I took it like a man, and quietly answered questions people were asking me. I overheard one woman say, "He is so brave!" I was pretty pleased that I made that impression. Eventually, I was able to get up and another player took my spot as backstop. As I ran out to left field, the crowd gave me a round of applause. I don't know if the incident had unnerved the other team or not, but we ended up winning the game.

It was a busy spring. Harry and his wife had four small children under the age of six. Harry was running a dairy operation and farming his one full and one quarter sections of land. Despite being a busy man, he was trying to help his wife with the babies whenever he could. Dad started working for him almost fulltime, and between Dad and me, we also got a crop planted at our place.

I started working for another neighbour, Mr. Brown. School had just ended, and Mr. Brown and I were working the summer fallow and making hay on his farm. He had just bought a brand new Massey Ferguson 65 diesel tractor. That was the most beautiful tractor I'd ever seen, and I loved driving it. I even own one now and still love to drive it.

I had only been working there for about two weeks when the Massey Ferguson dealer from Glenella came to see Dad. He had a small business that he and his uncle had handled up till now, but it was expanding and he needed a good mechanic. He made an offer that Mother couldn't refuse, and Dad took a job in Glenella.

When I came home from Mr. Brown's, my dad told me what he had done. He said, "You're going to have to take my place at Harry's."

I replied, "Mr. Brown isn't going to be happy."

He said, "I'll go and talk to Mr. Brown tomorrow. You start with Harry."

And that's what happened. I worked for Harry every summer after that for the next five years.

Harry didn't have a nice Massey Ferguson tractor like Mr. Cole or Mr. Brown. Instead, his family had bought a model U Minneapolis Moline in 1954, and an older one bought the year before I started working for him. I think it was a '48. The newer U was nice to drive, but that old one was a dog. It didn't steer very well, and it had high rear tires that were rougher, but it had a rebuilt engine that was excellent. I think I spent half the summer that first year on that old tractor, breaking sod with a ten-foot Massey Ferguson deep tiller hooked to the back. That old tractor would slip and then lunge forward, all the while bouncing over clumps of sod. It's only saving grace was you could stand on the platform and hang onto the steering wheel for dear life. There was no sitting down as it was far too rough a ride. Fourteen hours on that thing was a long day, but I did it till the job was done.

Working for Harry made up for it all. Mr. Brown was pretty critical, and I didn't like working for him too much, but Harry was an entirely different matter. When I started, he would have been thirty years old and liked to talk. Whenever we were together doing something, he was always talking. Not only did he talk to me, but he listened to me as well. I will always give him the credit for me turning out as well as I did. I learned later that every child needs to have a significant adult in their life. I was most fortunate that I had two—my grandfather and Harry. Harry appreciated my work, gave me responsibilities probably beyond my years, and trusted me with his farm and his equipment.

Whenever he had to spend time in the house with his wife and help look after the children, Harry left me to take care of the work outside. He would get up at four in the morning, go outside, and start the chores, milking the cows and feeding the pigs. He came in at six, got me up, and we had breakfast together. I lived there all summer; it was more convenient and I preferred it to being at home. I even stayed on into the fall, went to school from there, and helped with the harvest.

We planned the day over breakfast, and I loved that time with Harry. He often let me decide how some jobs were going to be done. Working for him helped me develop my potential and confidence. Then I would help him finish the chores and head out to the fields. I was in at lunch for a big meal, then back out to the fields till four in the afternoon. We came in from whatever work was going on and did chores together. We ate supper at six and went back out to the fields until dark. I got the lights working on the old U and would often come back to the yard after dark with the lights on.

I remember one weekend, on a Sunday, when Harry's wife asked me if I wanted to go home to see my parents. I responded I'd rather stay there because there was nothing at home for me. I was thirteen at the time. I was not bitter—I said it in a matter-of-fact way—but I think she was shocked. I offered to go out and work that day in the fields, but Harry and his wife were devout Catholics and told me I could just relax and have fun.

There were a lot of gophers in the pasture where the dairy cattle were, and Harry was worried that some of the cattle might break a leg in a gopher hole. So I took Harry's dog and a couple of pails of water up to the pasture. I poured water down the holes, and eventually the gophers came to the surface. Harry's dog was pretty quick as he grabbed those gophers and killed them.

When we finished that job, there was a pail of peas for me to shell. I was sitting doing that when the dog came over and lay down beside me, so I started feeding him the shells. He liked the attention, so he sat there beside me and, with some encouragement, ate all of those shells that afternoon. I believe the poor dog stayed in the bush for two days after that and wouldn't look at me for a week. Harry asked me what was wrong with the dog, and I said he ate too many gophers.

***

ANDREW COLE IS HARRY'S COUSIN. LAST YEAR when I went to Andrew's to hunt deer, I asked how Harry was doing. Harry was in his mid-eighties by then, living in a seniors' place in Neepawa with Alzheimer's. Andrew suggested we go see him, so we did. When we walked in, I saw the thirty-year-old man I knew in that eighty-five-year-old body. I asked him if he knew me. With a little coaching, he said, "Young Bill!" We went right back to 1962 and made hay and forked manure all over again. I felt accepted and valued and loved, and I was ever so glad I had gone to see my old friend.

CHAPTER TWENTY-ONE

Playing Politics

THE CONSERVATIVES WON THE ELECTION IN THE fall of 2016. Our MLA Ralph Eichler was appointed the Minister of Agriculture. I told my councillor Stuart that we intended to meet with the minister, and he suggested that his brother, who was chair of the Pork Council, might also like to attend the meeting.

On October 17, the chair, the general manager of the Pork Council, and I met Ralph, who seemed interested in trying to help us with our problem. I asked him for his support to effect a change in The Planning Act. I wanted to have it changed so the RMs could be held accountable if they did not enforce a count of animals in a livestock operation and take action where circumstances required. The Chair of the Pork Council stated that his organization did not support illegal hog operations. The Minister indicated that he wanted to have a review done of the situation and that he would get back to us with the results. I wasn't sure what that meant, but I was to find out.

I gave him a year to complete the review, and on November 19, 2017, I sent him an email asking for a meeting with Young Joe and me; we wanted to find out where things were at with the hog barn. He responded at one o'clock the next morning and suggested that things were as normal as possible. He went on to say that he had received no complaints and "they are very much aware that they are being watched and monitored regularly." We knew by the smell that this was simply not the case.

Replying on behalf of our group, the Concerned Citizens of Big Island (CCBI), I asked the Minister who exactly was watching and monitoring this operation. I told him we had a letter from Muddy Woods stating they were refusing to count the number of pigs, and also that in discussions with them, they advised me that they had an agreement with the colony that the colony would self-monitor. I was hoping the Minister saw the irony in my comments. Unfortunately, although not surprising, I received no response to my remarks or to my request for a meeting.

After the NDP lost the provincial election, they had a leadership race. I was approached by one of the candidates for my support, and I took that opportunity to tell him about our problem. He said if he was elected, he would like to help out. After he was elected, I asked him for a meeting in March 2018.

I met him at the offices of the official Opposition leader in the Legislature. I walked in, and he was there with his Chief of Staff. I remarked to him how posh these offices were, and he responded, "Yeah, not too shabby!" I liked him. He seemed to be a down-to-earth person and easy to talk to, and I spent the next hour doing just that. He was not familiar with terms like animal units (AU) and so forth, so I explained those things as we went along. I gave him an overview of the situation from the beginning, what we had done, where we were now, and what I hoped to accomplish. When he asked why the people who were affected by the barn did not run for the rural municipality (RM) of Muddy Woods council, I responded that the majority of us didn't live in that RM, so we couldn't run for office nor select councillors or reeves in Muddy Woods.

Our conversation turned to politics. I told him of our MLA's involvement while he was in opposition. I talked about the lack of progress that I had achieved with the agriculture ministers I had met while the NDP was still in power. The leader indicated that he would let our

MLA know I had met with him. We knew the Conservative government was making changes to The Planning Act, and he talked about putting forward an amendment to the current legislation. We agreed that we thought enforcement of The Planning Act should be moved to a government department. When I got home, I asked Young Joe to organize a meeting with our MLA. Young Joe, my councillor, and I met with Ralph on April 16, and he promised to take our concerns to the Cabinet.

That spring, I went to the NDP convention in Brandon. It was a great weekend. We took our camper and camped there with friends who were also attending. When I snagged the leader and brought him up to speed on our current situation, he told me there were hearings planned for mid-May with regards to Bill 19, The Planning Amendment Act. Because we wanted to propose some changes, such as to transfer enforcement of The Planning Act from the RM to the province, he suggested I get on the speakers' list and present to the standing committee.

I met the NDP Environment Critic at the convention, and talked at length with him about the situation. He was interested in seeing the operation for himself and came out to my farm on May 4. We toured the area, and discussed the concerns with the hog barn, aquifer, and surface water contamination related to manure spreading. I showed him the historic Grant's Lake Wildlife Management Area that was adjacent to the spreading fields and only a mile from the manure storage facility. He was very sympathetic.

On May 15, I presented to the Standing Committee for Bill 19. In my presentation, I provided a brief review of events leading up to the present, including the effect it had had on Dorothy and me and our relationship with the colony. I went into my most recent conversations with Eichler, who was also present. The MLA from The Pas, who was also the NDP Agriculture Critic, jumped on that and made some appropriate and pointed remarks that the Minister hadn't done anything. My MLA's face was usually quite red, but that evening it was almost purple. The truth hurts. Afterwards, I connected with some people from Hog Watch, who had been quite helpful to us over the years. All in all, it was a very satisfying evening.

At the end of May, we received a letter from the Minister's office suggesting that the RM of Snow Valley send a delegation to a council meeting at Muddy Woods. The letter said the RM should voice their concerns about the facility and the impact it was having on their citizens, and request that Muddy Woods conduct an inspection to ensure compliance.

And so we had come full circle. CCBI had gone to a Muddy Woods Council meeting several times and asked for a count. The two adjacent RMs had also gone to that council and asked for a count. One councillor from each of our municipalities—Snow Valley and Greenland—had gone to the colony and asked for a count. Now the Minister was suggesting we should do it again. If something doesn't work, are you supposed to keep trying the same thing over and over? It seems to me that that's the definition of insanity. Our provincial government was clearly just trying to pass the buck.

Nevertheless, we decided we would try once again and follow the Minister's suggestion. Snow Valley sent my councillor and several others as a delegation to Muddy Woods. On September 17, my RM of Snow Valley sent me a copy of a letter they had received from Muddy Woods, which stated that they had an agreement with the colony that included a format and schedule for AU monitoring and that no additional AU reporting would be required. On October 18, Snow Valley requested the details of the aforementioned agreement, but before a response came, there were municipal elections. My councillor did not run, so now we had a new councillor.

The reeve of Muddy Woods also did not run, and a new reeve was elected in a very close race. The woman who was elected was a person I had worked with closely in the past, and I considered her a good friend. She was a parent in the school where I was principal when I retired, and in my last three years there, we had developed and implemented a preschool program for four-year-olds. She was the parent representative on the steering committee, and kept the program going after I retired. It was an excellent program, and we were all very proud of it. If it had not been for her efforts, the program would have been cancelled long before it was. I called her up to congratulate her on her victory.

She knew about the struggle with the hog barn and was agreeable to a meeting with me to discuss it. I suggested that it might be best for her to meet with me after she had been sworn in as reeve. I was concerned that she might get herself in some trouble with the council if we met before that. I waited till after she was sworn in and asked for a meeting, but at that point, she refused to meet with me.

Later, I found out that her advisor was the former economic development officer who had given the colony the building permit for the barn in the first place. I thought she was an honourable and decent person. So much for that idea. I realized she is easily influenced and listening to bad advice.

Eventually, however, I would have a chance to express my views to her.

CHAPTER TWENTY-TWO

Leaving the Land

ONE SUNDAY IN THE SUMMER OF '63, we went down to Killarney to a family gathering and dinner at my dad's youngest sister's place. His brother was visiting from Ontario with his wife and daughter; they had married in England when he was overseas. He usually tried to come out to Manitoba every summer to visit with his family.

Dad really wanted to see him, but Mother was not anxious to be around her sisters-in-law. She felt they did not accept her, especially the English girl. The journey down there was a litany of complaints about how stuffy these people were. As usual, my father was a smoking volcano behind the wheel as he drove along at thirty-five miles an hour to keep the motor on the same old car from blowing apart. By that time, Marlene and I were taking books to read on these trips and didn't pay a lot of attention. I was happy because I really wanted to see my grandfather.

When we got there, Father and his brother went off for a walk to talk alone. Father seemed to be able to talk freely to his family, and he hadn't seen his brother for several years. Mother had a new baby to show off, and she was the hit of the afternoon. Grandfather wanted to know what I was doing, so he and I sat down and talked together in a swing on the porch of the house. That was such a nice afternoon and I felt really good spending the time with him.

Then it was time for dinner. We all sat around the big dining room table my aunt had set up, and enjoyed an amazing meal of roast turkey with all the fixings. I was becoming known as a big eater, and I did not disappoint anyone. I didn't get a chance to talk to my uncle much with Dad around, and I never really got to know him. I was never included in their conversations, nor was anyone else, including Grandfather.

When the event was over and we were leaving, Grandfather and I walked out of the house together with our arms around each other. People noticed and made some snide remarks about our relationship, which we ignored. He put his hand on my right shoulder as we were approaching the car and turned me to face him. He looked at me and handed me a beautiful compass. His voice cracked a little as he said, "I want you to have this. It will be there to guide you when I'm no longer around."

I pull that compass out from time to time and look at it, especially when I have to make a hard decision. I talk about it and have it with me when I tell his story, "Tales of the Homestead." The bond I had with him has meant so much to me over the years. Thank goodness he was there when I needed him. Now when my grandson tells me that I am the best grandpa ever, I know exactly how he feels.

***

IT WAS ANOTHER BUSY SUMMER WORKING FOR Harry. I was doing the mowing and raking while he looked after the chores and helped his wife with the babies. Harry had a Farmhand loader on the newer U, and would use it to sweep up a big bunch of hay and bring it over to the stack. I would sort it out a bit and pack it down, and by that time, he would be back with another sweep load of hay that he put on top of the first load. Then I would sort that and pack it down as well, before shaping a top with the last several loads. Harry came from each end of the stack with the loader, so we were building them two sweeps long and one sweep wide, with the hay piled to about fifteen feet high.

It was a fast way to make hay in the summertime, but gathering the hay and bringing it home in winter was more of a problem. Harry had a grapple on the sweep with which he could take a bunch of hay off the stack and put it on the rack. That sort of worked if the snow wasn't too deep and the stack wasn't frozen. Otherwise, you would have to take a hay knife, cut the stack, and load it by hand. Some people used stack movers to bring the hay into the yard, but Harry never bought one of those, although he was talking about changing to a baler that summer.

I loved building those stacks, and there's nothing like the smell of fresh hay. Hay in stacks is generally of a very good quality, and usually better than baled hay because it cures better. I feed sheep in the wintertime, and I have to say, there's nothing nicer than breaking open a bale of fragrant hay when it's thirty below. And the sheep think so, too.

***

IT WAS ABOUT THE MIDDLE OF AUGUST, one of the few weekends in the summer when I came home to spend the day on the farm. My parents took me aside and told me that we were moving. Dad didn't want to drive to work anymore, so they had rented a house in Glenella. We would be moving at the end of the month, and I would be attending high school there.

I was devastated. I sobbed and sobbed; my heart was broken. I loved that farm. I loved farming, and I loved the life that I had built in the district. Father promised that he would not be selling the farm, and told me he expected me to continue to run the farm in the summertime. That was some consolation. Harry seemed to understand how I felt and told me he wanted me to continue to come and work for him.

Father sold off what little livestock we had left, and then sold the horses. Fortunately I wasn't there when Trigger and Buttermilk left, and I was relieved to find out they went to a good home. I don't think I helped Dad move the family—at least, I don't remember doing that—but I believe some of the neighbours did. I was at Harry's and glad to be there while all that was going on. Mother was ecstatic. That farm had come to symbolize everything that was wrong in her life. She hated it with a passion and couldn't wait to leave.

I remember the evening that the neighbours came over to say goodbye. They were genuinely sorry to see us go. For all those years, my parents had successfully maintained the façade in the community that everything was okay in our home. I remember each of my parents getting up to speak and telling these people how sorry they were to be leaving. For Dad, it may have been true, but it certainly wasn't for Mother. My little sister Jean was as cute as a button, and Marlene and I said very little; I was glad not to say anything because my emotions were too strong, and I would have broken down. Harry came up and talked to me; he knew how upset I was to be leaving. Mr. Cole, Mr. Carter, and Mr. Taylor all spoke to me at some point during the evening. I appreciated that, but I think the best thing of all was a letter I received from my grandfather.

In that letter, Grandfather talked about his disappointments in life. He had been training to become a steam engineer in the British Navy. He was quite an athlete and, among other things, played rugby. In fact, he was trying out for the British Navy rugby team when he was badly injured in a game. He was on his way to making the winning touchdown when the opposing team piled up on him, leaving both his knees broken. He was not only out of the game, but also out of the Navy. He went home where his mother nursed him back to health, and within a year, he immigrated to Canada. He said it turned out to be the best thing that could have happened, and he would never have had it any other way. That was comforting. Thank you, Grandfather. Always there when I needed you.

***

AS I WORKED THE SUMMER FALLOW ONE last time on the farm, I pulled up some bones with the cultivator. There were some leg bones and a skull with a neat little bullet hole. I had found Sport's remains. I gathered up what I could find of my old friend and buried him under a tree where we had often rested together on a hot summer afternoon. And then we moved.

Bill's grandfather on his 80th birthday.

CHAPTER TWENTY-THREE

Back into the Fray

AT THE END OF NOVEMBER 2018, I received a copy of a letter sent to the municipality (RM) of Snow Valley by the RM of Muddy Woods. In that letter, the new reeve indicated that there was no formal or written agreement in place with the colony in terms of counting the pigs; the previous council had led us to believe there was one. The new reeve also indicated that the colony had been conducting and submitting counts to the RM every six months, and since 2013 while this was happening, had been within ten percent of the allowable number of animal units (AU). We knew this was not true as we had the results of a count that exceeded their limit in a Manitoba Farm Industry Board (MFIB) report in 2014.

On December 17, I called a meeting of the Concerned Citizens of Big Island (CCBI). We held it at the home of two young people who had just moved into the district and were within a mile of the barn. They had experienced the smell that summer and were interested in joining our group. One of our members picked up Harold's wife, who was still supportive of us, and brought her to the meeting. About a dozen people were there from the community, and the NDP Environment Critic came as well.

I was so pleased by the community support, and very proud of our group; we were and still are as strong and committed as ever. People naturally come and go in the community, but there is still a solid core. I have made some good friends because of it and am making more as time goes on. One of the perks of this job.

I went over the events that had happened in the past year, and then I asked the group for an endorsement of my leadership. Both the RM and the colony had tried to pin the problem on me, saying that it was just me complaining about the barn and nobody else. This was in spite of the fact that every time we went to a meeting with either the RM or the colony, the community was well represented. Of course, they were trying to isolate me and minimize the involvement of the community. It was a good tactic, but it didn't work.

The people there wholeheartedly indicated their support for my efforts and endorsed my leadership. We passed a motion that CCBI would request an audience with RM of Muddy Woods and demand action on the issues. The motion was passed unanimously. We decided to draft a petition that would be circulated in the community and also put online. The MLA volunteered to do that and also to organize a media event before we did our presentation to Muddy Woods. He would also help us craft a press release. I was to request a meeting with Muddy Woods and draft the presentation with another member of the group. We were on our way.

The MLA suggested we ask the Ombudsman's office to look into the matter of the change from 658 AUs in 2004 to 889 AUs in 2005, given that there weren't any motions, resolutions, or bylaws on the part of the RM that reflected that change. I think we were all very determined, and I was excited to be back in the fray. That evening, I filled in a complaint form to the Ombudsman and emailed a request on behalf of our group for an audience with the council. I used a quote at the meeting that night from William Shakespeare's Henry the Fifth: "Once more unto the breach, dear friends."

We decided that in conjunction with our press release, we would hold a public meeting at Big Island. In the press release we complained about the smell, concerns about water pollution, the refusal of Muddy

Woods to monitor the operation, and the failure of The Planning Act to protect rural residents.

On January 28, 2019, I made a presentation during the AGM at the Big Island Community Hall. I gave them a brief overview of events since 2004. I talked about our concerns with the colony falsifying information on their Manure Management Plans (MMP) and the long-term effects of extra phosphorus and nitrogen in the spreading fields. I reported that we had asked the RM to do a count of the animals in this operation as soon as possible and require the colony to comply with the number determined to be correct by the Ombudsman's office. I said that we wanted the province to do monthly air quality testing and conduct a full environmental audit of the colony's operation. I also spoke of our request that the RM cover the costs of the immediate testing of all domestic private wells supplying water to local residents adjacent to the operation.

I left them copies of the petition and asked them to get signatures. The petition asked for an independent third party count, the RM to cover all costs of private well testing, and the province to do monthly air quality testing and conduct a full environmental audit on the operation. I told them about the press conference at the Manitoba Legislative Building the next day—a first for me—at which the online version of our petition would be launched. And I informed them that on February 4 at seven o'clock in the evening, we would be holding a public meeting in the Big Island Community Hall to discuss this situation in-depth with residents. The next day a number of us travelled to the Legislative Building to request that the provincial government help solve the water and air pollution problems affecting our community. We all met in the NDP Caucus room and discussed what was about to happen.

At the appointed time, we went down to the rotunda of the building where a number of news outlets were present, including CTV. I talked about the unacceptable smell degrading our quality of life as rural Manitobans. We expressed our concerns about water pollution and harm to the historic Grant's Lake Wildlife Management Area and Lake

Winnipeg. We wondered why no one in government appeared to be interested in enforcing existing laws to protect the quality of life for rural residents and their communities. The Environment Critic talked about other operations in Manitoba that had illegally expanded their operation without the required permits or government approval. He complained that this government does not apply the law fairly across the province. I didn't bother to mention that we had had the same problem when the NDP was in power.

It was quite the event, and I actually really enjoyed it. Some of our members felt that I had perhaps gone too far in my criticism of the RM and the colony, but I didn't hear of any repercussions. The news clip appeared on CTV that night, and although some of our group thought the feature was silly and somewhat disrespectful to us in their presentation of our concerns about the smell, I felt we had got at least some of our message across. I was of the opinion that any publicity was good for us and bad for the colony, and was quite happy about how it had gone.

***

ABOUT A WEEK BEFORE OUR SCHEDULED PUBLIC meeting, I sent out a newsletter to the community, inviting everyone to come. A newsletter also went to the colony. I talked about our request for clarification about the number of animals at the colony and that we'd only had marginal success in getting the information. I mentioned that the adjoining two RMs had also requested that information without success. I talked of our concerns about excessive smell and the over-application of manure on the surrounding fields, and the possible effects that could result. I stated that neither the previous provincial government nor the current one had been helpful, and that it was left to us to try and resolve the situation now and for the future. I asked for their support and mentioned the petition.

On February 4, the day of the meeting, I got a letter from a company that was sampling and analyzing soil from the fields at the colony and writing the MMP. I called the company and had a talk with the manager. He assured me that they did the proper tests on the fields and gave the colony their recommendations for the application of manure. He was concerned that if our allegations were correct, his company could lose their licence to do this work. I asked him if his company also supervised the application of the manure, and he indicated that work was done by a different company.

Of course, I already knew the answer to that question—it was another colony that had been hired to do the application. A member of our group had observed their quad tractor and applicator applying manure at two in the morning with a steady rain falling. If that colony was prepared to spread in those conditions, I very much doubted they would be concerned at all about the over-application of manure. I imagine application rates of manure in the rain could not be monitored and most of that effluent would likely be carried into adjacent waterways.

I had intended to go to the school that day to have thirty copies of my presentation made. It turned out to be a storm day, and school was closed. I got about eighteen copies of my booklet prepared before my printer ran out of ink. Dorothy and another member of our group had brought cookies and coffee. We went about half an hour early to set up the hall and put out the refreshments.

About forty people turned up, including a reporter from the *Stonewall Argus*. One of the first things I did was lift two large briefcases of material onto a shelf where everyone could see them. I said I had collected about forty pounds of paper over the past fifteen years in my efforts on their behalf. Thank goodness that I had kept a paper record because my computer had crashed the week before the meeting. Just before I started my presentation, three or four people from the colony came in and sat at the back of the room—neither Peter nor Sam attended. There were about four people from the community that sat with them and several people at the front that I knew would be supporting the colony.

I handed out my booklet that included an overview of everything that had gone on since 2004. I picked the highlights that I wanted to talk about in particular, and I went into some of those in more depth. I briefly mentioned some that I thought the community might find amusing, especially my interactions with Conservation and the provincial government. In my summary, I explained how many pigs an AU actually represents, since most people think the numbers refer to individual pigs. Then I mentioned our current complaint to the Ombudsman about the number of AUs the colony should actually have. I indicated that the RM of Muddy Woods insisted the colony was in compliance, despite the report by the MFIB that indicated the operation had 1,125 AUs during the same time frame. I told them there were no penalties for falsifying numbers on the MMP and that we had no idea what the nutrient levels in the fields actually were. I mentioned my conversations with the company that did the MMPs, as well as the fact that they had no idea what the application rates were. I spoke about our frustration with the enforcement of The Planning Act, and that under the present laws we were powerless. I explained that when I asked for the Act to be changed, the current government had weakened it even more. I mentioned our request to have all private wells tested in the area, and our concern about the manure storage facility, which was nearly twenty years old.

I felt good about my presentation and really enjoyed doing it. I also think people found it enlightening and perhaps even entertaining. Some members of the community complimented me on all the work I had done, which as you can guess, was music to my ears.

I concluded by asking them to sign a petition and invited them to attend the meeting that was coming up on February 12 with the Muddy Woods Council. I then opened the floor for questions.

Several of the colony's supporters at the front of the room immediately started with arguments that they had used back in 2004—basically that the new facility took care of the pollution problems created by the biotechs. This was certainly an old song and did not get much traction with the group. One of the people sitting at the back with the Hutterites asked me if I had trouble with my well. It came across as some kind of dig or sly attack, so I responded that my well was fine and I wanted to keep it that way.

That was the extent of the opposition, but there were many other questions for me. After the meeting, I spent some time with Adam Peleshaty, a reporter from the *Stonewall Argus*, and the article appeared in the edition of Thursday February 7.

We were generally pleased with how things had gone and impressed with the turnout. Young Joe expressed his opinion that I had put on quite the show. I was pleased with that and am looking forward to other opportunities to do it again.

CHAPTER TWENTY-FOUR

A Day of Reckoning

AT THE END OF AUGUST 1963, WE moved into an old two-storey house at the south end of Glenella. It had a big yard with room for a garden, and I even had my own bedroom. At the back of the house was the CNR main line that went up to Dauphin. The occasional passenger train went all the way to Churchill, but mostly it was big freights thundering by. I enjoyed watching them, despite the noise, and pretty soon I didn't even hear them at night when I was asleep.

We had taken Shep and Queenie with us, and they slept outside on the porch. Queenie had a bad habit of chasing vehicles, and she didn't last more than a month before she was run over and killed. I gathered her off the road, took her out to the farm, and buried her beside Sport. Shep adjusted well to town life. He was getting old and settled into retirement. He was always so glad to see me.

In September, I started Grade 9 at the high school. It was a small school with only three teachers, but there was a great feeling of camaraderie. I had the opportunity to take British history, and on our first test, I got ninety-nine percent. That was unheard of in that school, and news of my accomplishment made the rounds in the town.

There was one student in my class who had repeated a year and was a bit of a bully. I don't think he liked the fact that I was doing so well academically. I remember one day when he tried to hurt me, and he ended up being stretched across a desk in a very uncomfortable position. Some Grade 12 students were witness to this event and told him that he had met his match. I had no trouble with him after that and made a number of friends right away. I was glad to be attending school there.

After school, I would go over to the shop where Dad worked. I was eager to learn and wanted to be helpful, and the dealer had work for me. His yard was full of used tractors of all kinds, and I got to drive a good many of them. When a tractor came in on trade, he would send me out to his farm with a cultivator or some other piece of equipment to give it a workout; he wanted me to find out if there was anything wrong before he resold it.

Occasionally, I would borrow a tractor from him and go out to a nearby bush to cut wood. The farmer who owned the property let me use his four-wheeled wagon to bring it home. One tractor I used was an International A, and I was impressed with how much it could pull and how well it could handle the deep snow. Dad complained that the cost of gas he bought to put in this tractor did not cover the wood that I cut. When the dealer started paying me for my work, I had a little money to buy the gas myself.

There was always a new tractor on display—a Massey Ferguson 35, 65 or 90. The dealer did not have a truck and trailer, so when these tractors were sold, someone had to drive them out to the farmer and bring back the trade. That was often me. These days, I restore the same tractors that I used to drive around the country when they were brand new sixty years ago. I presently own a 35, a 65, and an 85, and use all of them on my farm. I have the same feeling of excitement when I drive them now as I did back then.

New tractors came to Neepawa and had to be driven the thirty miles to Glenella. I was always game for that; as long as I got a chance to drive a new tractor, I didn't care if it was night or day or even how cold it was. I learned how to change tires, including the big rear tractor tires, and the older men who worked in the shop really appreciated my doing that job. In the wintertime, there were always two or three tractors being rebuilt, and Dad was teaching me basic mechanics while I helped him with these overhauls. This was a public place, so he treated me better when we were there.

I was becoming quite a valuable worker, and the dealer offered me a job for the summer. I told him I was going back to work for Harry. I explained that my first love was farming, and when I worked at Harry's, I was also close by our farm and could do the work there as well. What I didn't tell him was that it was mostly because I wanted to continue to develop my relationship with Harry.

***

MY GRANDFATHER CAME TO LIVE WITH US that winter. He was getting older, and his little house in Ninette was both cold and had no running water. We didn't have running water either, and I don't remember him ever taking a bath. He always had a pleasant smell about him that reminded me of old wood and tobacco, a smell that to this day is comforting for me. That winter, I would go to school for the day, go over to the shop and work for several hours, and come home with Dad for supper. After supper, Grandfather and I would sit and talk and play cribbage.

He told me many stories of his childhood and his time in the British Navy. He had loved his mother dearly and remembered her well. He talked about his exploits in school, which I found interesting and amusing, and he wanted to know what was happening for me at school and at work, and usually had something funny or wise to say about it. He was proud of me, and I loved him very much. He was an intelligent, well-read man who knew the most amazing things.

He would try to get my father to talk with us or play cribbage, but Dad usually sat in the corner with his face in a newspaper. I don't think I ever really saw him interact meaningfully with Grandfather that entire winter. He did not seem to want to be part of anything that the two of us were doing. I, on the other hand, was so glad that my grandfather had come to spend the winter with us because it was our opportunity to get to know and appreciate each other, and we made the most of it. Surprisingly, Mother and Grandfather got along well. He was someone who would talk with her, and she liked him.

That spring, I bought a tractor with the money I had earned working in the shop. It was a Leader, had a Hercules engine in it, and a three-point hitch. It had a top speed of about seven miles an hour with the three-speed transmission. I didn't realize it at the time, but it was quite a rare tractor, and I didn't see another one until a neighbour in Big Island brought one over for me to work on about fifteen years ago.

Now that I had wheels, I could drive the tractor out to the farm when Dad refused to take me. It was thirteen miles from Glenella, and took me a couple of hours, but it was better than walking, which I had done several times. Dad was funny that way; I'm not sure why, but I think it was because he wasn't that energetic himself, and when I worked in the shop, I got more done than him. I know that bothered him because the dealer and others would compare us as workers. Instead of being proud of me, he was threatened by that.

It all came to a head when I had just turned fifteen. I wanted him to drive me to the farm to do something, and he said it wasn't necessary to do that work. I had farmed for several years with Harry, who was an excellent farmer and had taught me the correct way of doing things. I don't know what I said, but suddenly Dad attacked me. I saw him coming and turned to face him. I don't think he realized how strong and wiry I had become; changing tires does a great deal for your muscle tone. I wrestled him to the floor. I had his arms twisted behind him, one knee on his back, and the other on the side of his head. He struggled, frothed at the mouth, and swore at me. I knew that I was pressing his face into the spittle on the floor, and I didn't care. I felt all of the rage and hurt and fear that had built up in me over the years, and it gave me the strength of ten. I was totally in control, and he was totally out of control. I felt really powerful, mature and responsible.

I'm not sure how long he struggled, but he did his best. Finally, he exhausted himself and stopped struggling and swearing. I let him up, but he could hardly stand; he was finished and panting for breath. I was in much better shape and still able and willing to defend myself if I had to. He didn't come at me so I started to say something conciliatory, but he refused to engage, turned, and staggered away. I wondered for a moment if he was going to have a heart attack.

I watched him go, and felt my heart turn cold toward him. Any slight feelings or respect that I might have held for him disappeared in that instant. He never touched me again although he certainly used lots of sarcasm and disdain with me. At that point, I really didn't care, and I'd curse him to his face every opportunity I got, but only in private. I had learned from the best how to keep quiet about family secrets.

When I look back on it now, I'm surprised he didn't kick me out, but then that would become common knowledge in the community. I think that was the only reason he didn't. I was perfectly capable by then of making my own way. I was a good student and a good worker, and I could have paid for my own keep and still gone to school.

Mother found a new use for me after that incident with Dad. She realized I could protect her from him, and I was to end up doing that for the next forty years. He was still having his uncontrollable, silent, violent rages from time to time, and she would threaten him with telling me if he directed them at her.

I usually only heard about these episodes, like the time she had gone for a walk one night and locked the door to the house behind her. When he came home while she was away, he thought she had locked him out and kicked the door in. I heard about that when I came there the next weekend and found the damaged door. By then I was boarding in the little town where I was teaching.

***

HARRY HAD TRADED OFF THE TWO U Minneapolis Moline tractors that spring and bought a Minneapolis Moline M5 gas. It was such a nice tractor to work with. When I brought the Leader over, Harry put gas in it and I used it for raking. It had no water pump, and on a hot day, it would start to boil. When the summer was over, I traded it off and used the money I had earned to buy a Minneapolis Moline model R. Now that was a tractor! That winter, I overhauled it in the shop, and the next spring it was ready to go. It had a road gear of about twelve miles an hour, and I drove it regularly to the farm and to Harry's. It pulled the seed drill, harrows, and a six-foot one-way. I took the R to Harry's to do the mowing and raking.

When the summer was over and I had a little bit of money, I traded off the R for a Massey-Harris 30. The 30 had a five-speed transmission, which was better suited to haying, and was an excellent tractor that had more power than the R. But ever since I've regretted that foolish decision to let go of the R because it was a great little tractor. I still have the 30 in the back and promised myself that one day I will restore it.

Harry started having troubles with the M5. The power steering was leaking, and he tried to have it fixed several times with no success. Eventually, the M5 started giving us lots of trouble, and when it wouldn't start, I was able to use the 30 to pull it. Harry was getting pretty frustrated with it, so I wasn't at all surprised one Monday morning when I came back to work to see that the M5 was gone and there was an International Harvester 560 diesel in its place. I liked that tractor. It wasn't as pleasant to drive as the M5, but it was a workhorse and very reliable.

I had experienced many tractors up till that point, and I was about to move into the world of cars.

CHAPTER TWENTY-FIVE

Shaking Up the Council

AFTER THE PUBLIC MEETING ON FEBRUARY 4, 2019, the two local papers—the *Stonewall Argus* and the *Stonewall Tribune*—covered the event in great detail. I thought both papers did a fairly good job of describing the meeting and discussing the situation. The reeve of Muddy Woods talked extensively in the second article and quoted a number of statements from the Department of Agriculture, which I suspected were not accurate. We had been talking to that Department ourselves, so we knew the slant the reeve was giving their comments was not wholly correct. It was also possible the Department was telling the RM one thing and us another.

The one thing she said that really made my blood boil was that, based on her conversations with Department officials, the council would not be counting the pigs. She had made her decision and stated it clearly in the local paper before we were even able to make our presentation to the council.

When I looked at the individual fairness checklist the Ombudsman's office had prepared for municipalities, it clearly asked the question if the person affected by the decision had been given an opportunity to state or present their case. This was obviously a violation of the correct code of practice. I filled in another complaint form for the Ombudsman, and in this complaint, we accused the rural municipality (RM) of

Muddy Woods of partiality towards the colony against the community. I mentioned a number of occasions in the past where we felt that the council had been biased.

On February 12, we attended a Muddy Woods Council meeting. I'm not sure how many people they expected, but seventeen of us showed up. I was invited to sit in one of the large comfortable chairs at the table with them. I declined, saying I would rather stand with my friends and neighbours; it was a statement that held a great deal of significance for me. I then turned to one of the more elderly members of our group and invited them to sit down instead of me, a gesture they sincerely appreciated.

I started out by introducing the people from our group to the council; they did not return the courtesy. Before the meeting, I had emailed the CAO a chronology of events that had transpired since 2004. I asked if they'd seen it, and they said they had. We told them we were not anti-Hutterite, nor were we against the hog industry. We were simply requesting that the colony obey the law and the RM enforce it. We knew the colony had been over their allowable limits on three different occasions. And we were concerned about the three sets of numbers of animal units (AU) that we had seen since 2004, as well as the fact that there were no bylaws, resolutions, or permits in municipal records that reflected those changes. We told them we had placed the matter in the hands of the Ombudsman.

The council had claimed that the colony had been in compliance since 2013, which we knew was not correct because we had a Manitoba Farm Industry Board report from 2014 that stated they had 1,125 AUs. When I said that to the council, they claimed they had never seen the report, so I gave them a copy. I then brought up the article in the February 7 edition of the *Stonewall Tribune* in which the reeve had stated she would not have the herd counted; I reminded them that this was without even listening to our presentation or hearing our arguments.

At this point, I got a bit animated in my tone and volume and realized my standing position allowed me to dominate the room, with the support of the Concerned Citizens of Big Island (CCBI) behind me. By the look on the reeve's face and her posture, I knew she was feeling pretty uncomfortable. I told her we had also taken up our concerns about her apparent impartiality issues with the Ombudsman's office.

We wanted to know if they could tell us where the excess manure was going, conduct impartial testing of well water, and support us in a request to the government of Manitoba to conduct air quality testing and a full environmental audit of the colony's operation. I mentioned that the last time I presented to the council I had requested that a committee be formed to solve this issue, and I was making that request again. The meeting reached a crescendo with a number of us asking the RM what they were prepared to do and when they were going to help us. The CAO tried to keep other people from speaking, but I said there were lots of people who had things to say, and after that, he didn't try to squelch them.

When someone asked whether the council would take on the responsibilities that they were assigned under The Planning Act, we were met with silence. We asked a number of other questions with the same result. The reeve asked us if we were finished, and I stated we wanted answers to our questions and we would be back to hear them. When we left the office, I scheduled the next meeting for March 26.

We gathered in the vestibule, and we were pumped. Several of the women gave me hugs and high fives. And while some of the group thought I may have been a little too intense, most of us certainly felt good about how we had conducted ourselves after all those years of frustration.

On February 21, a great article appeared in the *Stonewall Tribune*. The article talked about our group being forced by the behaviour of the council to take our case to the Manitoba Ombudsman. It explored our complaints in depth and covered most of our arguments. It also stated that the RM had not responded to the paper's request for a statement.

Our MLA Ralph Eichler, the Minister of Agriculture, was quoted in the local paper: "Communities work best when they are working together, so I encourage the two sides to keep the lines of communication open." I emailed him and told him we couldn't agree with him more and that we had practically pleaded with Muddy Woods to work with us. I urged him to use all the powers at his disposal to help make that happen. He had also stated in the article that "... various government departments have been instructed to address and resolve this unfortunate long standing dispute."

I had known the Minister for years, long before he became a politician, and addressed him on a first name basis. I knew he was quite religious, so I asked him to take a verse from his own hymnal, referring to his comment in the paper about keeping the lines of communication open. I wanted him to tell us what he and the various government departments had done and were doing in this situation.

I also wanted to talk to him about the changes I was proposing to the standing committee on Bill 19, The Planning Amendment Act. I said that because this great country believes in the rule of law, a situation such as this should never exist. That a group of people, such as ourselves, can have such an injustice foisted upon them, and then be totally unable to rely on any of their governments for protection if they choose not to enforce the law, was very wrong. I told him that a large group of us had been at the meeting with Muddy Woods, and that many of us were wondering what he'd be doing about this matter and where he suggested we go from here. Perhaps we could organize a meeting with him at Big Island. I posed a number of questions to him, and he responded that he would pass them on to governmental officials that could answer them. He put us in contact with a woman in the Department of Agriculture who was very helpful and seemed to honestly answer the questions we posed.

On March 1, we received a letter from the RM. They told us they were working with representatives from Manitoba Agriculture and the municipal lawyer to review the information available. They claimed that the RM had been upholding all its legal obligations to the colony and the surrounding citizens. They assured us that they valued concerns such as the importance of environmental controls, but that such matters were a provincial jurisdiction. They knew we were in contact with Manitoba Agriculture and assured us that they were committed to ensuring that lawful municipal practices were being followed for the benefit and safety of all citizens, including CCBI and the colony. They promised us an update once they had concluded their investigation.

We decided we would wait and see, but expected little. And as it turns out, we were wise to do so.

CHAPTER TWENTY-SIX

Airborne and Amphibious

THE WINTER OF 1964/65, I WORKED AT the shop almost every day after school and on Saturdays. I had saved up a bit of money and was turning sixteen in May. On my birthday, I walked into the municipal office after school and asked the secretary to sell me my driver's licence. She asked if I could drive, and before I could respond, said, "Oh, yes, I've seen you driving around town with the dealer's truck." And that was that; she gave me my driver's licence.

Two years before, when I was fourteen, I had driven Harry's 1946 two-ton Ford grain truck to Kelwood with a load of flax. Harry was a little reluctant to send me with that old truck because the flax was valuable. He had no place to store it, and didn't wish to stop combining to take it himself. The old truck had non-directional surplus army tires of about three different sizes, and indeed, I was halfway to Kelwood when one inside back dual blew. The truck settled down a little bit on the outside dual, but fortunately it held, and I was able to make it to Kelwood. I unloaded the flax at the elevator and brought back a cheque for about five hundred dollars to Harry. That was a lot of money back in 1963.

The next time I drove that old truck, he had put better tires on the back. I loved driving that thing with its straight-cut gears and unsynchronized transmission. There was a real art to shifting because you had to juggle the engine speed and double clutch to get it into the gear you wanted without grinding. I had also driven the dealer's 1958 three-quarter-ton Dodge pickup a number of times, but usually on back roads. I had to come into town right past the municipal office, and that's when the secretary had seen me driving.

That spring, my dad decided to buy an old John Deere D. The G was getting pretty shaky, and we needed a different field tractor. My 30 wasn't powerful enough to pull the cultivator, although I used it for many other jobs. The D had a straight pipe out the side and no muffler, and I blame it to this day for some loss of hearing in my left ear.

I didn't want Dad to buy it. I'd driven many other tractors, especially the new Masseys, and that old D was a real beast. I told him to hold out for a Massey 44 or a Minneapolis Moline U—anything but that D. He bought it anyway because the dealer let him have it for $250, and he wasn't planning on driving it much at all; that was my job. It did start and run reasonably well, but it had a tendency to overheat; those tractors also didn't have a water pump. My dad told me to slow down or stop for a while if it started to boil, but I would deliberately overwork it and get it boiling, hoping that it would quit and he would have to buy another tractor. Of course, it never did break down, and I was stuck with that thing.

It was a distillate burner, and distillate was no longer available, so we would mix gas and diesel fuel together. It didn't work all that well because after a while the fuels would separate, and you would be burning either straight diesel or straight gas. It also didn't have hydraulics, so Dad bought a hand crank to use in place of a cylinder on the nine-foot Case deep tiller he bought with the tractor. It actually worked well because you didn't have to lift the cultivator at the end of the row. You just turned with the cultivator in the ground. That meant the headlands at each end of the field were always well worked. I didn't like the tractor, but I was happy with the job it did.

Mr. Cole came into the shop one day and complained about the Massey 88 gas he had bought from the dealer. Andrew still owns and operates that tractor today, and they have always had trouble with it. Mr. Cole had traded his John Deere D for the Massey and was saying quite loudly that he wished he had his old D back. Because I had just taken our D out to the farm, I told him that he should take that Massey over to our place, leave it there, take our old D home, and everyone would be happy. Mr. Cole's face turned red, and he couldn't even think of a comeback. My remarks generated considerable mirth at the time in the place, although poor Mr. Cole was quite embarrassed. Needless to say, that trade never happened, but I would have been delighted to have that Massey 88.

The dealer put a lot of pressure on me that summer to come and work in the shop, but I was steadfast because I wanted to farm and spend the summer with Harry. I also knew my dad was happier when I was not there. I had become a valuable worker at the dealership and some of the farmers were coming to me when they wanted things repaired. My dad usually came across friendly, but as you worked with him for a while you realized things were not quite right; he could often be sullen and silent. He was never a happy man.

I know there was some tension between him and the dealer, and that the dealer valued me more than he did my dad. He would often choose to work with me, and usually I would go on service calls with him. Later on, when I graduated high school, he wanted me to become the foreman in the shop. I refused because I knew as long as Dad was there, that would never work. Besides, there was a whole exciting world out there waiting for me.

I was very busy with all sorts of work that summer, and got to know the International Harvester 560 very well. It still wasn't the most comfortable thing to drive because it had no power steering and was a coarser tractor than the M5, but it was powerful, reliable, and rugged.

At the end of that summer, I thought I had enough money to buy and register a car. Dad didn't want me to buy a car and figured I wouldn't have enough money to even put gas in it. But I was determined, so we went to a garage in another town and found a very nice 1955 Pontiac, teal coloured, four doors, six cylinders, and three-on-the-tree—a three-speed manual transmission with the shift lever on the steering column. The only problem was it had a slight knock in the engine, but I found that if I added a couple of cans of STP when I changed the oil, the knock went away. So I handed over the $125 he wanted for it, and I had a car.

That fall, the movie *Psycho* came out. It was playing in Kelwood, and some of the girls in Glenella suggested that I take them in my car to see the movie. They hadn't wanted anything to do with me up to that point, so I said no. I wasn't that desperate to keep company with them if all they wanted was a ride. And I was not interested in horror anyway because I had seen enough of it at home growing up.

It was shortly after that I decided to take my car out to see what it could really do. I was travelling down a municipal road at about eighty-five miles an hour when I came to a crossroad. It didn't occur to me that these roads were crowned—graded higher in the middle—and there was quite a ridge where I was about to cross. I was upon it before I realized what was happening, and I just had time to take my foot off the gas before the Pontiac crossed that intersection and became airborne.

When it came down on the road again with a crash, the car filled with dust. It started to swerve this way and that, and I couldn't hold it on the road. I turned off into a ditch, and there was a huge boulder half the size of the car on the side of the ditch. I think I might have scratched the car a little bit as I slid by!

I swerved back up on the road, and there in front of me was a parked car. Two faces, one male and one female, popped up as I went by. One of the faces belonged to one of the girls who had wanted a ride to the movie. I guess the flying leap the Pontiac took had thrown some oil out of the engine because it was knocking as I drove by and waved. I went home and added another couple of cans of STP, and it was right as rain again. I parked the car so the scratch wasn't visible when Dad got home. That girl had trouble facing me when we met at school, but I didn't say anything to anyone about it.

The Grassy River meandered through the country just north of Glenella. The road ran beside it and curved here and there, following the river and making for a winding scenic drive. I was coming back home along it one afternoon when I noticed the gas pedal had fallen off. It had a nasty habit of doing that, and I had fixed it several times before. But when I bent down to put it back in place, I suddenly felt the car go over the edge of the road and into the river. A wall of water came up over the hood and the windshield, but I could just see a little island in the middle of the river straight ahead. I steered for it, and when the car finally stopped, it was perched on that little island. It was about four o'clock in the afternoon and I considered my options, none of which seemed good.

I decided to walk back to the farm—about thirteen miles along the correction line—to get the 30 so I could pull myself out. I ran-walked to the farm and got there about six o'clock. I started the 30, grabbed a bunch of chain, and headed back to the car. It took a little over an hour to get there. I laid out the chain, hooked up the car, and attached the other end to the drawbar of the tractor. The little tractor did its best, but it just couldn't pull that car out of the river!

By that point, it was about seven-thirty, and I drove back to the farm. There was still daylight, so I started the D and headed out at full speed—five and a half miles an hour, pukka pukka pukka pukka—down the road. It was about ten o'clock when I got there and hooked onto the car. The D was a heavy tractor, and it managed to pull the car out of the Grassy River. I left the car on the side of the road and headed back—pukka pukka pukka pukka—to the farm. By the time I got there, it was about two in the morning. I parked the tractor and started to walk back to the car, arriving about four-thirty. The car had dried out by that time, so I started it, drove home, and got into the house at about five in the morning.

That morning was Saturday, and I was supposed to be at work at the shop by eight. Dad woke me at seven. He asked me what time I got in the night before, and I told him I wasn't sure, but I thought it was maybe midnight or something. When we left for work, I hoped that he wouldn't see the bulrushes that were caught underneath the Pontiac.

About ten-thirty, Dad went over to the café for coffee, and when he came back, he said one of the people who lived along the Grassy had seen a teal-coloured Pontiac in the river the previous afternoon. I told him I didn't know anything about it, and there were lots of those kinds of cars around. He went to the café again about three o'clock for afternoon coffee, and when he got back, he had this smirk on his face. He said somebody who lived along the correction line had heard a John Deere tractor going by at about one o'clock that morning. I knew the gig was up, and told him, "Yeah, okay, you got me."

That old car lasted me almost two more years until eventually the engine sort of exploded. There is still a piston out in Harry's field somewhere! I guess I'm lucky I didn't kill myself—not then, nor as a result of some of my driving antics during those years. When my son turned sixteen and drove recklessly, I couldn't say too much, even when he smashed up one of my cars. In fact, I helped him stay out of trouble for that accident, something he appreciates to this day.

Last year I told him about the Pontiac because I thought it was time for him to know. Although his son is only eight, I can see what is to come by the way he drives the model car with a lawn mower engine I built for him. I'm giving the boy lessons in my little standard Toyota Echo on the farm. It's time, and he is very interested. This way I have a chance to teach him properly.

That fall, I was old enough to go deer hunting. I had bought a beautiful little .303 Mark 5 Lee Enfield army surplus rifle, and Dad and I drove down to his father's farm in Ninette to hunt. It was as if he had gone back to 1940 before he enlisted in the army; he was in those hills again as a young man. This was a side of my father that I had seldom seen, except perhaps when he was with his family of origin. He was full of stories about hunting deer in those hills, and it was as if the War had never happened.

I was the one that got a deer that day, and for once, he was pleased with me. When deer season rolled around each year, I wanted to go hunting with him, as he did with me. We went back to Ninette several times, and even out to the homestead near Lonely Lake in Manitoba where Grandfather had lived back in 1918. The last time Dad hunted with me he was about seventy-five years old. He finally quit because the cold bothered him too much.

I still get out every season, and now I prefer to hunt by myself. When I'm out there, I always think about those good times, and it seems I am never alone. I believe he has made sure of that. I love my time hunting deer.

CHAPTER TWENTY-SEVEN

Pig Headedness

I REALISED THAT A NEWSLETTER IS AN excellent way to communicate with the community at large. The colony hated the newsletters because they do not like to be in the public eye. Still, I needed to tell the community that we had two investigations going on with the Ombudsman regarding Muddy Woods.

I sent out a newsletter that described what those complaints were and promised to keep the community informed on developments. I mentioned our group's commitment to continue to meet with Muddy Woods until they came up with tangible plans of action to deal with the situation. I let them know I had decided to ask for all the Manure Management Plans (MMP) from February 2014 until the present. I explained I was interested in seeing if the colony would reveal any information to us, since they had to give permission for Manitoba Sustainable Development to provide that information.

We put out another news release that outlined our complaints to the Ombudsman and the upcoming meeting with Muddy Woods on the twenty-sixth. We also received answers from the Department of Agriculture to a series of questions we had posed to the official that the MLA had put in contact with us. Other than finding out what specific regulations relate to Conditional Use Hearings, we didn't really learn anything new.

We set up an appointment to talk to another lawyer, and three of us went to speak with him. Our first lawyer had done some work for Muddy Woods, so working for us again would have put him in a conflict of interest. The meeting was essentially to describe the situation and what we had done up to the current time. When the lawyer agreed to meet with us, I sent him a seventeen-page document with basic information about our situation. We had a little trouble finding his office because he had recently moved, but during the meeting, he seemed to be interested in our situation and quite knowledgeable about municipal law. He even agreed to write a letter to the rural municipality (RM) of Muddy Woods on our behalf.

When March 26, 2019, rolled around, most of us were back to meet with the council. I was planning to stand again to make my presentation, but this time the room was crammed with so many chairs there were more than enough for my group to sit down. The reeve told me to sit, and I responded, "I do better on my feet!" She then *ordered* me to sit down, and I looked at her and decided this wasn't a hill I wanted to die on. When I sat down, there was a collective sigh in the room. I'm not sure if it was because I had complied, or because the reeve had lost her opportunity to throw me out.

We had twenty minutes, and she read to us a two-page letter they had prepared to address the concerns we raised at the first meeting. She was nervous and had difficulty with some of the language. That took up fifteen minutes of our time, which seemed quite deliberate on the part of the council. There were no surprises in the letter and some obvious misconceptions. For example, they stated they had no legal right to require a third party to count unless substantial evidence is obtained. That leads one to wonder what evidence is required and how it can be obtained. Our view was that the smell was evidence enough to trigger an investigation.

When they said they could not obtain the MMPs, we knew that was completely wrong. They disputed our claim that the colony had built the barn too big for it to be considered a replacement for existing facilities, and indicated it didn't matter anyway. They concluded by stating that the remainders of our concerns were environmental and beyond the jurisdiction of the RM.

Then it was my turn.

I asked the council to introduce themselves because they had not done so at either the beginning of the current meeting or the previous one. I then started with a statement that even though the colony had promised the community not to expand their operation, we had proof they had not only broken that promise three times, but continue to do so. And we offered to pay for the initial count if the colony turned out to be in compliance, otherwise the colony would be responsible for costs. We asked the council how they wanted to proceed—mediation perhaps, or other ideas that came to mind. We assured them that we are responsible and concerned members of our community and not activists. We said we'd been asking for and willing to work with the council and the colony for years, and asked if they would now agree to count the pigs. We told them it was the belief of the Minister and the Pork Council that the barn should never even have been built in that location, and I reminded them that they had permitted it anyhow. We asked if any research had been done on the size of the barn required before a permit was approved, even though we already knew the answer to that question. We talked about the misinformation we'd been provided over the years and how that had destroyed the trust between the RM, the colony, and ourselves. We talked about the lack of opportunities for development, the concerns about people wanting to buy property in the area, and the effect on property values for the rest of us. We reminded them that they had indicated they would fulfill their legal obligations and demonstrate lawful municipal practice in their last letter, and to us that meant they would count the pigs.

At the end of the meeting, a number of us heard the reeve mutter, "If we count the pigs, will you go away?"

I responded with an enthusiastic yes, and a number of our members whispered to me, "Conditions, Bill, conditions!" I quickly added I'd send them a list of our conditions, which I did as soon as I got home.

***

I REMEMBERED SOMETHING THAT MY COUNCILLOR HAD said about the colony being religious and that they did not seem to be overburdened with compassion for the surrounding community. I contacted the third brother, Mark, in the colony, and asked if I could sit down and talk with him about what the colony was doing to the community. He said he had nothing to do with the pig operation and that I needed to talk to Sam and Peter. I told him I thought that he, as the community's religious leader, should be concerned about the way the colony was treating the community. I got no response from him in terms of the meeting, so I put my thoughts down in a letter and mailed it to him. I didn't receive any response to that either.

***

INFORMATION FROM THE DEPARTMENT OF AGRICULTURE STATED that in 2004, the colony had a 1,250 sow-to-weanling operation that was also finishing some pigs, and that the constructed manure storage facility had a capacity for a 1,250 sow, farrow-to-finish operation. Therefore, the province recognized that the pig operation was in fact a 1,250 farrow-to-finish operation and they were finishing pigs on the colony yard. What they didn't know is that while it was true the colony had a 1,250 sow-to-weanling operation before expansion, it was also true they had been finishing pigs off-site in several locations, but neither the RM nor the province knew anything about that. We knew because the colony told us at the community meeting back in 2004; they also said that the expansion would allow them to bring all their pigs home. We found out that one of the sites they were using was near Libau, Manitoba, after one of the members of CCBI saw the colony unloading pigs into a barn at that location. The province had incorrectly assumed that the entire operation was being carried on at the colony when in fact it wasn't. I suspected the RM was aware of the provincial position in this matter, but I don't know if the council knew there were pigs being finished off-site.

I sent out another newsletter to the community reporting on our meeting on March 26. We stated that the council still refused to count the pigs, citing legal reasons that we knew were incorrect: (i) the RM has no legal right to require third party, or even first party, counts—that was completely wrong; and (ii) the RM does not receive copies of the MMPs that are protected by the Freedom of Information and Protection of Privacy Act—maybe not, but they can request them. We mentioned that the RM had received incorrect information from the Department of Agriculture about the size of the operation on the colony before the proposed expansion in 2004. The MMP that the farm boss had filled out in July 2004 indicated there were 1,250 sows farrow-to-nursery and 2,000 feeders at the colony, making a total of 658 animal units (AU). We reported that the RM receives an email from the colony twice a year in which the colony itself reports the number of AUs in the hog operation. We told the community that after the reeve asked if we would go away if they agreed to count the pigs, we had sent the RM a number of conditions that we felt would be important for a count to be accurate. I then invited any community members who could to attend the next meeting with the council, which was scheduled for April 23. And I also asked my cohorts, "Do you think the council is quite tired of us by now?"

One councillor who seemed upset with my treatment of the reeve at the first meeting was not present for our delegations after that. I wondered how she could carry out her responsibilities if she wasn't at the meetings. In any case, at the next meeting, we were to get some very interesting information.

CHAPTER TWENTY-EIGHT

Moving On

WHEN I TURNED EIGHTEEN, I WORKED FOR Harry all summer and then went out to the west coast with my aunt. I had wanted to get a job on a fishing boat, but the season was over and there were no jobs. I took an aptitude test at Canada Manpower and the results suggested I should get into social work or teaching. Since Manitoba was so short of teachers that *they* were paying the tuition, I decided to go into teaching. I also believed it would have the added benefit of allowing me to farm on the side. It was the best decision I ever made and have never regretted it.

I pumped gas in Vancouver that winter and made arrangements to attend teachers' college in Brandon, Manitoba, the next fall. When I came back to Manitoba in the spring, I got a job driving a fuel truck for the local British North America fuel dealer. On weekends, I worked in the shop and saved as much money as I could to take the one year education course that winter. The next summer, after I had completed the course, I came back to Glenella and worked in the shop all summer.

In the fall of 1969, I started my teaching career—a grade 5-6 class in Brookdale Elementary School.

I taught in that school for five years, and in 1974, I resigned and moved to Winnipeg. I married a girl I had met in Glenella; her father owned the fuel truck I had driven. I worked for a year as a parts technician in the Co-op farm equipment company, and in 1975 I accepted a teaching position in a school division just north of Winnipeg. My wife was a city girl, but adapted well to farming, learned to drive a tractor, and helped me on the farm. I had made an arrangement with Dad to rent it for the next ten years after I got back from B.C. At the end of that time, he wanted to rent it to Marlene's husband so I bought a quarter section in eastern Manitoba and moved my equipment there.

That was a busy time for me. Besides changing farms, I became principal at the school where I had been teaching, and my son was born. A little girl was born to us four years later. My brother-in-law looked after the home farm for two years until he and Dad had a falling out.

By this time, Dad had been working down in Morden for about twelve years. He retired in 1982 and went back to farming. I think those retirement years were the happiest of his life since the War. He bought a John Deere R and some other equipment at a sale. He loved going to auction sales, and he hauled things home to the farm by the truckload. He was cutting wood out there and taking it back to Morden to burn in his wood stove. We had missed some years of deer hunting during that time, and he started hunting with me again. My farm in eastern Manitoba had plenty of deer.

The R lasted for a number of years until the hydraulic pump went. Dad replaced it with two tractors—a Case 730 diesel, and a Deutz 65. The Case was good, but the Deutz had something wrong with its governor that regulated the speed of the engine. Turns out the people who sold it to him knew about it, but didn't tell Dad. He brought that tractor home, but it just sat there with the old R. The old D was still there, and he bought a nice John Deere 60 with a loader. I told him if he ever wanted to get rid of that tractor, I would buy it from him. He had a fifteen-foot versatile swather and a small model C Gleaner combine. He was in business.

Mother had found a job in Morden and refused to go to the farm. Dad would spend four or five days on the farm every week, do some work, go to sales, and visit the neighbours. He made a good friend up at McCreary and would often go there for supper and spend the night.

I would drive out to Kelwood from time to time to check on him and see how things were going.

This went on for about ten years, but about the time he stopped deer hunting, he also stopped farming. He was finding it difficult to climb up on the tractor because of a bad hip and arthritis. I tried to talk him into a hip replacement, but he would have no part of that. Senselessly stubborn. He rented the farm to a neighbour and continued to go there to spend time and grow a small garden. By this time, the Case 730 had developed a leaking head gasket, and he started talking about getting rid of the equipment.

In 1992, following the breakup of my first marriage, I sold the farm in eastern Manitoba and bought this place at Big Island.

***

THAT WINTER I MET DOROTHY AND THE next summer, we got together. She came out to the farm at Kelwood with me and together, we cleaned up Dad's cottage—a small house my first wife and I had moved to the farm. There were a number of things that I thought should be thrown out and put them in the back of my truck. When Dorothy and I were busy in the house washing walls and floors, I would see Dad go and collect most of the stuff out of the back of my truck and hide it somewhere. He liked Dorothy, and she was good to him, even though he wouldn't remember her name or her connection to me.

When he started to seriously talk about getting rid of some of the equipment, I told him I would buy the Gleaner combine. So one day Dorothy and I went out there, filled it with gas, and headed for Big Island. It was a 120-mile trip at ten miles an hour—12 hours! I drove by the south end of the lake where there were tons of mosquitoes. Lucky for us, I discovered that because they couldn't fly quite as fast as I was going, the mosquitos sat in a little cloud just behind the grain tank—behind me. That little combine ran beautifully all the way home.

Before we left, I reminded Dad that I wanted to buy the John Deere 60 from him. He mentioned that he had promised it to his friend at

McCreary who was also going to take the R and the D. He wanted me to take the 730 Case because he knew I could fix it. He gave the Deutz to Marlene's husband, probably because he knew it didn't work, and indeed my brother-in-law couldn't fix it. Dad still held a grudge about what had happened between them when he had been renting the farm. Marlene's husband was making hay there for his dairy cattle and Dad wanted him to bale up a very weedy patch. He refused because the weeds would have tainted the milk. Dad got angry, and that was that.

I gave Dad what he wanted for the 730 and drove it home as well. It took about ten hours for that trip. I replaced the head gaskets and sold it to a farmer near Glenella. He came for it and drove it back again. I did make a little money on that deal.

The next summer, Dorothy and I got married. One of the parents from my school who was a minister came and performed the ceremony. We bought nice gifts for my, now our, two children, and there were lots of kids from both sides of the family that celebrated with us. Danny's Whole Hog catered the dinner. A steam locomotive for the Prairie Dog Central chugged up the tracks, and we had our wedding pictures taken in front of it. We had a band, there was dancing, people camped in our yard, and everyone had a great time. As if on cue, the sheep broke out of their enclosure, and everybody took pictures of Dorothy in her wedding gown chasing the sheep back in.

For some reason, my dad made a big deal about his present for me. He gave me an old bag of baler twine, which was fine, because I had a baler and could use it. But he gave it in a demeaning and sarcastic way, putting me down in the process, which suggested that was all my marriage or me was worth. That hurt, and I was quite angry about it. I didn't say much at the time, but I thought about it over the next week, until I decided how I was going to deal with it.

Dad had a habit of putting me down, especially talking about things I had done in my childhood, and I was sick of it. I went out to his friend's place at McCreary to talk to him about the John Deere 60. I told him that I had asked Dad for it, but he said he had promised it to him. Dad's friend was a decent guy, and he told me he had not realized I wanted it. He told me he'd paid $450 for it, and when I suggested that I could give him that amount of money if he would part with the tractor, he agreed. He also said that I should have had the first opportunity to buy it because that's what families were about, wasn't it? I didn't bother to tell him that wasn't the case in our family because I knew he could see that for himself.

I brought the tractor back to Big Island, and I didn't tell Dad a thing. I waited for him to show up at our place, which I knew would happen eventually. Sure enough, he showed up about a month later. Revenge is best tasted cold! As he drove on the yard, I could see him staring at the tractor, which I had parked conspicuously on the driveway. Of course, he recognized it right away. I didn't say anything, and I waited for him to initiate a conversation about it.

It didn't take long.

"That's my tractor," he said.

"Yes," I replied.

"Why is it here?" he asked.

"I went out to your friend's place and told him I wanted it. He said he didn't know that I had asked for it and family should come first. I gave him the money for it that he gave you."

Dad didn't know what to say. I could tell he was discombobulated about the whole thing. Mother was gleeful because she hated this friend anyway. She was always suspicious of Dad's friends and thought they meant more to him than she did, which was probably true. Dad made some excuse and was out of there like a shot. He had to take Mother back to Morden because she wouldn't go to the farm, and then he headed up to talk to his friend.

About a year later, I ran into his friend when he was visiting Dad in Morden. I asked him how things had gone, and it appeared as if neither the friend nor my dad quite knew what to do about the situation, and both were somewhat embarrassed by it. I quite liked the man and had several good conversations with him after that happened. He told me he and his wife really liked old people and enjoyed having my dad around. I was quite happy for Dad that he had these people in his life. Mother, on the other hand, would spit on the ground at even the mention of his name. I eventually put another $1,500 into that tractor and sold it for $3,700.

After that incident, Dad seemed to be more careful with what he said to me, and things got better. I wished I had done something like that years earlier.

***

DAD'S BEEN GONE FOR FOURTEEN YEARS NOW, but I continue to go out to Kelwood and hunt on Andrew's farm. A couple of years ago, I went out with my camper and it turned out to be an extremely cold weekend. Andrew invited me to come and stay with him in the old farmhouse where he had grown up; I knew that place well as a child. Andrew is by himself there, so it was just the two of us together that weekend. As you would expect, we fell to talking about our families, and I told Andrew some of the things that had gone on when I was growing up. It felt good to share my experiences with my friend.

If you remember, the Coles were our neighbours and farmed just across the road from us. I knew that Mr. Cole had seen some things or heard things that went on at our farm. Andrew was also discreet, like his dad, but I had opened the topic and it was just the two of us. He told me about a couple of situations that happened when Dad was about seventy-five years old. He had been working in his yard when Dad came over and asked Andrew to help pull out his little pickup that had got stuck on the back road leading to the farm. Andrew said he was seething. They went back to the truck with Andrew's tractor, Andrew backed up to the truck, and hooked up the chain. He told Dad not to do anything, just steer the truck, because his tractor was more than capable of pulling it out. But Father wouldn't listen. He was so angry that he revved the truck, put it in gear, and when it was clear of the mud, smashed into the back of Andrew's tractor. The tractor wasn't damaged, but the front of Dad's truck was pretty messed up.

At this point, Andrew said to me, "I'm going to tell you something that I've never told anyone." One Sunday afternoon, he was working the field across from our farm when he saw a vanload of Hutterites driving up our road and into our yard. Andrew didn't think much of it because he knew Dad was there. Suddenly, he saw the Hutterite van come tearing out of the farm down the road in a cloud of dust, bouncing through the badger holes. In that cloud of dust was Dad in his little pickup, hanging out the driver's window, pointing a rifle at the van. Both vehicles disappeared down the road.

I was not totally surprised because I had heard rumours from another friend out there that something like that had happened. Apparently one of the Hutterites had said something to him. Where there is smoke, there is fire.

Bill's dad on the farm at age 80. Drawn by local artist Linda Gillies.

CHAPTER TWENTY-NINE

Department Hanky-Panky

ON APRIL 22, 2019, OUR LAWYER WROTE an excellent letter to the rural municipality (RM) of Muddy Woods. He stated that the colony was approved for 658 animal units (AU) and had self-reported that number on its 2004 Manure Management Plan (MMP). The increase in intensity to 889 AUs had not been permitted, and the original capacity of 658 AUs should have been retained. He indicated that the RM was incorrect in their position that they had no powers of inspection. He quoted the section of The Planning Act that permitted enforcement. He also indicated that the doubling in the permitted size of the barn was unnecessary and, as such, was not a replacement of existing infrastructure. This could be viewed as an intensification of use by the colony and was not permitted under The Planning Act.

It was our position that the RM was not upholding the zoning bylaw in this case. The RM seemed willing to believe the colony's assurances that they were in compliance, in spite of the three occasions when they were proved not to be. Was it because most of us were not residents of that RM and they did not feel responsible to our community? Some people in Big Island had, on more than one occasion, made it clear to me that they never had a very high opinion of the communities north of them; they considered them to be comparatively poor and less intelligent. I deeply resented that because I knew lots of good people in those communities. That led me to speculate that perhaps councillors from Muddy Woods may have seen this as their chance to get even, but I kept those thoughts to myself. Our lawyer concluded by saying that we had attempted to work with the RM and the colony for many years in relation to this substantial community issue presented by the barn and the operations of the colony.

The next day, we met with the council. As we were waiting for our delegation to begin, three men from the colony walked in, one of whom I vaguely recognized. "Is that you, Sam?" I hadn't seen him in twelve years, and his beard and hair had turned white. I asked him if Jacob was still alive, and although I never really got an answer, I did get the idea he still was.

When it was time for our presentation, we went in and sat down. I confirmed that they had received my correspondence about conditions for doing a count after the last meeting. I reminded them that the council had acknowledged there were no additional bylaws or resolutions that affected the change in the number of AUs, so the number was not approved as the council had stated in their letter read out at the meeting on March 26.

I then spoke about the provincial assessment that the colony did not overbuild their facility. It was clear from the 2004 MMP that the colony had a 1,250 sow-to-weanling operation, which was 375 AUs, and they were finishing approximately 2,000 animals, which was 283 AUs in ten biotechs. This added up to 658 AUs in total. Obviously, the province was misinformed about the operation on the colony at that time. We knew that the colony was using a barn at Libau, Manitoba, and other sites to house feeders and finishers. I was attempting to make it clear to the RM that the Department of Agriculture had based their findings on wrong information about the operation on the colony back in 2004.

I said it was apparent that the council was not in possession of accurate facts or even had an understanding of the size and operation of this facility. We strongly recommended the council create a subcommittee to deal with this issue. That committee could obtain all pertinent information, consult with expert witnesses, and have representation from ourselves, the RMs, and the colony. At the end of my presentation, I had planned to ask the council why there were no representatives from the colony at these meetings. Instead, I turned to Sam and asked what he had to say. The CAO jumped in right away, and said to Sam, "You don't have to answer any questions." I immediately responded that he should have the opportunity to speak if he wanted to, just as the members of our group had the right to speak if they wanted to.

Sam launched into a criticism of our concerns about spreading manure. I brought him back to the topic at hand by telling him it was the number of pigs and the manure they produced that concerned us. He didn't say much after that. One of the councillors spoke up and, in spite of being warned by the reeve, said he felt the council was between a rock and a hard place and that the province was urging them to allow the colony to have 1,250 AUs. I was ready for this and told him that I had talked to a Department of Agriculture official who had assured me that the province would never advise the RM to break their own bylaws. Privately, though, I wasn't so sure. I wouldn't put it past some of those officials. There does not seem to be a correlation between what they do and what they say.

I reported all this to the Ombudsman, and they said they would let me know if they can investigate this allegation. The RM said they had received our lawyer's letter that morning and would respond to it in due course. Again, we gathered in the vestibule to debrief, and the Hutterite men hurried past us in silence, not meeting our glances.

Because we were not expecting cooperation from the RM, we went back to see our lawyer to ask what further options we might consider. He said before we could consider suing, we would need to launch a complaint in writing to the Manitoba Farm Industry Board (MFIB), and he offered to help us with the complaint. He felt we had excellent grounds and should be successful. He also suggested we might want to approach Manitoba Sustainable Development to discuss any potential investigations that could take place related to the hog barn operation.

I made a request to meet with Muddy Woods again on May 28, but the council refused our request. They implied that we had no additional new information to share with them. I put together an email that listed the different topics we had covered over the three meetings and indicated there was little repetition. I also stated that we had brought forward new research in this matter on each occasion, which would be true of our next presentation as well. I sent a copy to the Minister of Agriculture, but did not follow up because I was contacting the Ombudsman to restart their investigation. I simply wanted him to know what was going on and that we were suspending activities for the moment while the investigation was underway.

On May 2, a very good article appeared in the *Stonewall Tribune* that summarized our difficulties with the RM and the colony over the years, as well as recent developments. I was then approached by several people who said the article had helped them understand our situation, and they were very supportive. I contacted an old friend in Hog Watch and told her what was happening. She said she might be able to get me a copy of a letter that had been written to another member of Hog Watch. That person had complained about an operation in another part of the province, and Sustainable Development had gone in and counted the pigs. Eventually, I got a copy of that letter, which I intend to use to urge the provincial government to also do a count for us.

I then received some more correspondence from the Department of Agriculture in which they admitted that their assessment was not based on a point-in-time count of the pigs. They still insisted they recognized the pig operation as a 1,250 sow, farrow-to-finish operation, based on what they had heard when it was developing. They refused to acknowledge that they had been misled. Just another example of how the laws are twisted to the advantage of the hog producer, and how these government officials refuse to admit when they are wrong, even when it is pointed out to them.

I received the MMPs from 2014 to February 2019 that I had requested a month before. The MMP for February indicated 741 AUs in the operation. I believe the colony allowed us to see that number to make us think they were in compliance. However, I already knew it was wrong, and that confirmed for me they were misrepresenting the numbers on their MMPs.

On May 23, we sent out a newsletter to the community. We reported on our discussions with the lawyer and our April 23 meeting with the RM. We said that the council would not schedule a delegation for our group, so we had come to the conclusion that the only substantial information the RM would accept would be a count, and the only body that could order a count was the RM. We reported that the Ombudsman had told us they had suspended their investigation until Muddy Woods' response to our lawyer had been submitted. And we told them we were considering formal complaints to both the MFIB and Sustainable Development.

Along with the newsletter, I also sent out a smell chart and asked community members to keep track of the smell over the summer. I had developed that chart as an easy way for people to keep track. It included information about date, time, temperature, wind direction, and humidity. I hoped consensus in that information from everyone affected might make a difference in some way at some point.

***

ON JUNE 5, WE RECEIVED MUDDY WOODS' response to our lawyer. It said that the RM had been provided with the AU information from DGH Engineering in 2004, and had accepted this information as the operating level. They claimed that the province advised that the primary purpose for the MMP is to ensure livestock manure is stored, handled, and disposed of in an environmentally sound manner. That begged the question, why would they even want to know the number of AUs if that information is not important? They did agree that the RM has broad authority under The Planning Act to administer and enforce its zoning bylaw and carry out inspections. However, they felt that the measures they were already taking were sufficient.

Our lawyer's comments on the letter indicated he did not see any obvious errors in the position stated. He indicated that he thought a court would be reluctant to interfere with the RM's discretion to conduct an investigation. We talked to him about the possibility that the RM was in contravention of The Planning Act because of the size of the barn, and he was prepared to investigate that further if we wanted him to. There was also the matter of money and how much we wanted to invest in this possibility.

Theresa, one of our members, went to a Hog Watch meeting in Portage la Prairie on June 14. Her comment to me afterwards was that Hog Watch should organize a coordinated approach to address concerns with the hog industry in Manitoba.

On June 18, our group met to discuss what had been happening. We'd asked a local real estate agent to come and talk to us about the effect that this hog operation might have on our property values. He was the parent of a student in one of the schools where I had been a principal, and was quite candid in his remarks. He said that he personally would not sell a property within two miles of an intensive livestock operation because of the effect that it might have on the future possibility of sale and the dissatisfaction of the buyer. He also discussed concerns for nitrogen in the aquifer and phosphorus pollution on the surface and the possible contamination of private water supplies. One of our neighbours had high nitrogen levels in their drinking water, but put it down to another source. Nobody else in our group has that problem that I know of—at least, not yet. He didn't tell us anything we didn't already know, but it was good to hear someone in the business confirm our concerns.

We talked about the letter Muddy Woods had sent to our lawyer, and his remarks about it. Theresa reported on the Hog Watch meeting she attended and her thoughts on a coordinated approach, and I talked about a conversation with the CAO of the Pork Council. I also handed out a smell chart that I had previously sent to the community, and asked the group to record the data.

The member of our group who hosted the meeting had put on a great spread, which we all enjoyed before taking our leave.

***

AT THE BEGINNING OF JULY, I MADE a private connection with someone at Muddy Woods and we had a conversation about the situation. They wished to remain anonymous, but suggested that someone other than me should approach the colony and ask for a meeting. When my contact indicated the colony was considering arbitration with us, Young Joe and another member of our group said they would be willing to give it a try. I passed that information back to my contact and asked them to let the colony know we were interested. I hoped this meant that perhaps the nut was beginning to crack.

CHAPTER THIRTY

The Experience of a Lifetime

WHILE MY SON WAS IN THE NAVY, I bought him a twenty-seven-foot tender from the Canadian Navy. After his tour was over in 2010, he and I started to develop it into a cabin cruiser. I think this idea came to me when Dad started talking about building a forty-foot boat that he was going to sail to England. He even went so far as to cut down a big poplar in the bush on the farm for the keel. Except for me, the whole family derided him for this idea, scoffed at him, and said it could never be done. I suggested that I would help in the building and sail with him. I knew very well how skilful Dad was and how determined he could be. Even though I supported him and his idea, I think he faced such ridicule from the rest of the family that he became discouraged. That big poplar lay in the yard on the farm for many years until I dragged the rotted remnants into the bush.

Dad was talking more and more about his war experiences and about revisiting England, France, and Holland where he had served. His brother-in-law, who was also a veteran, had gone over to England several times to visit the cousins who lived there, and had seen some of the battlefields in Europe. I think it was about 1993 that Dad decided that he was going to fly to England to revisit his experiences there for a couple of weeks. For some reason or other, he wanted my mother to go with him. She steadfastly refused, perhaps because she had never valued his contribution in the war, and also possibly because at that time she was threatened by his relationships with other people.

I told him I would gladly go and help him find his way around to see the things he wanted to see and reconnect with people, but he refused to take me with him. It was another occasion when he was unreasonably stubborn and suffered the consequences. He naïvely believed that the England he knew would still be there after fifty years.

After landing at Heathrow Airport, he stayed in his hotel room for a few days, recovering from jet lag and was unable to contact the people he wanted to see. After three days, he felt well enough to fly home, although he had planned to be there for at least several weeks. I was still working then, and told him that if he wanted to go the next summer, I would take him. But I think he was so shocked by how much England had changed that he didn't want to go back. Instead, he took me and his grandson on a fly-in fishing trip into Ontario where my son caught a master angler's pike. It hangs proudly in my basement.

Over the next few years, Dad pretty much quit farming. He had rented the land to a neighbour and just went out to stay in the cottage for three or four days and tend a small garden. He stopped cutting wood when that became too difficult for him. He and my mother were living in Morden, and every day would walk down to a local restaurant for coffee and a snack. About this time, he stopped doing that as well because of the arthritis in his hip, and when it became too difficult for him to get into a car, he started staying at home while Mother went out.

His hearing was going as well, so I suggested hearing aids. Not surprisingly, he would have none of it, although I think that was partly because then he didn't have to listen to Mother. Dorothy and I would drive down to Morden to see him. He liked Dorothy because she always cared for him in some way. I remember her down on her knees at one point, trimming his toenails. It was quite the job because they were thick and hard. I thought he might like to go out to see his farm at Kelwood and drove him there once, but he showed no interest in the place and wouldn't even get out of the car. Next, I tried taking him to his dad's farm at Ninette where we had hunted deer. He seemed more interested in going there; I could tell by the stories he told when we were on the way home.

When he was about eighty, Dad suffered a small stroke. He didn't seem to be affected physically, but became very emotional, and I remember him coming to visit me one day. Mother drove, and I could see it was a bit of a struggle for him to get himself out of the car. He wanted to talk about his war experiences. It was like that time when we had gone into the bush and he told me stories over the fire, but these stories were different; I had never heard them till that day.

He talked about the men with whom he had trained serving in the Sherman tanks as drivers and gunners. The Sherman was a poor tank with a gas engine, and because they caught fire easily, the Germans called them "Tommy Cookers." The tanks and trucks Dad drove all had radios, which they were instructed to leave on. Unfortunately, that meant he often had to listen to the screams of those young men, some of whom he knew, dying by fire.

Dad had been riding his motorcycle and guiding convoys over mine-strewn roads in France and Holland at night in 1944. When he contracted infectious hepatitis and was airlifted back to England, of the four guys that had gone over to England together, two had already died in tanks and the third one took over Dad's job on the motorcycle. Within two weeks, that man had driven over a mine and was killed. That haunted Dad for the rest of his life. He knew if he had been on that motorcycle, it would not have happened.

Dad was engaged to be married before he went overseas and corresponded with his fiancée all through the war. When he got back in 1946, they reconnected, and he wanted to get married, but not long after, she broke off the engagement, saying that he had changed. That corresponded with what my grandfather wrote once about Dad after he got back, describing him as bitter. Dad told me he had kept track of her all her life and she had just passed away. He was devastated. He still had the twelve hundred dollars that he had saved during the war for them to start their life together. That explained a lot of things for me. He had held a candle for her all his life.

It was around this time I contacted Veterans Affairs. The person I spoke to was surprised that there was a Second World War veteran around that he didn't know about. When I told him they had helped Dad with a loan many years ago, he was surprised they had lost track of him. It was obvious that my dad needed their services, and they were more than happy to help. And that was when my father finally got hearing aids.

***

HE WAS STILL DRIVING HIS LITTLE TRUCK, although he found it hard to push in the clutch because of his hip. One day, I got a call from him asking if I could help him get his truck out of the hills at Ninette. He had driven up there to my grandfather's farm in the bush, and the clutch had gone out. He had managed to walk with the help of his cane to the nearest farm about three-quarters of a mile away. He almost didn't make it.

When he phoned me from there, I went out to rescue him and his truck. Together, we went back into the hills with my four-wheel drive Dodge. I hooked onto his truck, and with him steering, pulled it out of those hills to a garage in Belmont where they could replace the clutch. He and I had cleaned out the box of his truck before we left and had taken his tools and other things back to Morden. When the truck was repaired and we went back to get it, he got very angry that his things were missing; he believed the garage had stolen them. At that point, I realized his mind was going.

The next time they went to the doctor, my mother told him about the incident, and the doctor took Dad's driving licence away. He was devastated. But then about a week later, she phoned me and said Dad was still driving that truck.

I went down there to talk to him. "Dad, you can't drive!"

He responded, "But I'm a good driver."

I told him, "You don't have a licence."

He said, "But I do have a licence," and pulled out his expired one. Before I left, I disabled his truck so it wouldn't start. That turned out to be a mere challenge for him, and because he was a better mechanic than me, it took him about fifteen minutes to fix.

About a week later, my mother called and said, "Your father is still driving that truck." I got the feeling that she was quite enjoying this.

This time my son and I went down there, and I said to Dad, "You know that truck needs a tune-up. Why don't you let us take it to a mechanic at Big Island, and we'll get it all fixed up for you." He agreed to that, and when we took the truck, we never brought it back. I had tricked it out of him.

I clearly remember the conversation I had with Dad on the phone several weeks later when he demanded that I bring his truck home. Finally, he said at the end of a heated discussion, "Well, we're not going to argue about this." And that was the end of that. I knew that another small piece of my dad died that day.

After he couldn't drive his truck anymore, he hardly left the house. As he became more and more reclusive, his friends stopped coming to see him, and he became isolated. I went down to see him one day in the middle of July, and he asked me if it had snowed that morning. He wasn't bathing or shaving, and the smell was pretty bad. I again contacted Veterans Affairs.

The first thing they suggested was for me to get power of attorney, which Dad agreed to immediately. I was actually more than a little surprised by that. Veterans Affairs was wonderful and provided anything I asked for. I had a ramp put on the house, and a room on the main floor converted to a bedroom. He still wanted to go upstairs, however, so I had to put a padlock on the door. The downside of that was when he stopped climbing those stairs, his physical condition deteriorated. I felt I had no choice because I was afraid he would fall.

Home care was provided, and when Dad was diagnosed with prostate cancer, a very expensive medication was covered. I told him what

Veterans Affairs was doing for him, and his response was he didn't think they should because they'd done enough for him already. Once all this was in place, he was not only quite grateful to them, but also a lot more comfortable.

I know he appreciated what I was organizing for him because one day he said to me, "You know, you were pretty hard to get along with back then, but I guess I was, too."

It wasn't much, but it's hard to describe the feelings I had as I drove home. It was such a profound relief and comfort that he had finally taken some responsibility for the way he had treated me. I was able to begin to forgive him then. After that, I would tousle his hair sometimes and really started to feel good about looking after him.

Of course, he could still be rude, like the time when I took him to the doctor, and while I was discussing his condition, he took out his hearing aids. The doctor noticed and commented on it, but I dismissed it, saying this was typical behaviour on his part. It was at that visit that the doctor said to me, "Your parents have an interesting relationship." I just rolled my eyes.

Those hearing aids were the second set that Veterans Affairs had bought for him for about $7,000. The first set had mysteriously gone missing about a week after he got them. I asked him where they were, and he wouldn't tell me. I eventually found them in the load of tools and other items from his truck after my son and I took Dad's truck and cleaned out the box. I think they worked well, but my guess is he couldn't stand listening to Mother and simply threw them in the back of the truck. When we got the second set, he resigned himself to wearing them. I made sure this time he knew their value.

My sister was going down to Morden once a week to clean the house, and her husband was cutting the grass and trimming the hedge. My parents had been treating her poorly ever since Dad had a falling out with her husband twenty odd years before. Dad seemed to have forgotten who she was and called her by his younger sister's name.

I remember Marlene telling me about one time when she was down on her hands and knees washing the kitchen floor, and my younger sister's daughter got down to help her. Both my mother and Jean laughed at the girl for doing that. That wasn't humorous, but disrespectful to Marlene. I'm still angry when I think about how Marlene was treated. Jean had married a guy who owned several factory hog barns and had three children. Once her children were born, Dad and Mother had less and less to do with mine. Marlene and I were also having less to do with her family.

Whenever I went down, I would buy groceries for them. Mother would give me an order, and I would go to the store and pick up whatever she asked for. Sometimes when it didn't seem to be very much, I bought about twice as much as she wanted. Whenever I got back, she was upset, and said, "Why did you buy all of this? We will never eat it in a month!" I just put it down to my mother being her usual miserable self.

My father started falling on the floor in the middle of the night when he would go to the bathroom. The first time it happened, Mother called the ambulance and the paramedics came, picked him up off the floor, and put him back in bed. After this happened several more times, I knew something needed to be done. Home Care would not come into the home during the night, so we had him panelled—the process necessary to put someone in a public personal care home. The decision was made for him to go into an assisted-living facility that was part of the Tabor home complex in Morden. He actually seemed to be okay with that decision until he realized that Mother was staying back in the house. I think he knew what she had been up to. I had not seen him that angry for a long, long time. More on this later.

If he had been physically capable, I know we would have seen one of his extremely violent episodes. At that point my anger for him blinded me and I did not fully realize what was going on. When he lost control of himself, I escorted Marlene out of the room and left him to the staff. He became very depressed and basically lay on his bed for a year and a half. Hindsight is 20-20, but I wish we had brought him home to end his days with us. Dorothy would have welcomed him with open arms.

There was an incident where he became very agitated in the facility and frightened the other residents. He was looking for the children that he thought had gone missing. I had heard his stories about Dutch children who were starving when the Canadians arrived. I think that was what it was about. The facility were going to make Dad leave, so I went down there to talk with him and the director, and we decided to give it another try.

A couple of months later, in late November, I got a call from one of the staff at about two in the morning. My father had disappeared. I asked her to check outside the building, and then I phoned the Morden police. I was almost dressed when the phone rang again; the staff member had found him on a bench just outside, in his night clothes, in below freezing temperatures. In his apartment, the bathroom door was beside the front door that led into the hallway, and his room was located right next to an exit door. I think he was going to the bathroom and ended up in the hallway instead. He turned to the left, which then led him out the exit. The door had panic hardware, and he was locked outside. I often wonder what he thought when he found himself out there. He loved the outdoors, and I think he would have happily perished outside in the cold. Unfortunately, it was not to be that way.

I went down to Morden the next day and met again with the director of the facility. We decided that it was time for him to go into a home where he could be safely locked in. He was eighty-seven.

When I told him he was going into the home, he calmly said, "I didn't think it would happen so soon."

I tried to distract him by saying, "Dad, do you remember us going deer hunting at Ninette?"

He simply responded, "I wish I was there now."

He was quite lucid, sad, and resigned that day. I felt so sorry for him that it had come to this.

I went down to see him from time to time, and he seemed content enough and was well cared for. After enlisting in the Navy, my son also went to visit before he was called up for basic training. He told my dad what was happening, and Dad responded that he hoped he could swim.

My mother stayed in the house for a while, and she came to see Dad with me occasionally, but I don't think she ever went on her own. That winter she slipped on some ice on the sidewalk and broke both arms. She couldn't look after herself, so after several days in the hospital I brought her home and we looked after her while her arms mended. Things got worse between her and me once Dad was out of the house. She then moved into the same facility that he was in, but into different quarters. She wouldn't have anything to do with him. That must have hurt dad, but she didn't seem to care.

***

WHEN WE CLEANED OUT THE HOUSE, MARLENE found copies of cashed cheques made out to our younger sister. We thought something was wrong, and then Marlene suggested it might be a form of elder abuse. Just as bad for us was the fact that both of our parents had kept this a secret from Marlene and me. Another example of their favouritism towards Jean and her family, and their disregard towards us and our loved ones.

Although I was supposed to be in charge of their finances at that time, I knew nothing about those cheques. I didn't think there was a need to keep *that* close a watch on things, and I had got power of attorney because it was required by Veterans Affairs. Marlene was quite angry and wanted me to do something about it right away. I decided I would wait until an opportunity presented itself. Sure enough, Mother called me one day with concerns that the Morden Credit Union was taking her money. That was my opportunity.

I went to the Credit Union and got a copy of the last year's cheques. They had all been signed by our parents, and the amounts indicated Jean had been taking half or more of their retirement income. When I confronted Mother with this information and explained that this was where her money had gone, she simply said, "Oh, yes." It was clear she knew all about it and had just forgotten. At first, she was prepared to do something about it. She said she wanted to make things right with me, but would not do anything for Marlene. I was having no part of that.

After several days, she reversed course and refused to do anything at all to correct the situation. When I went to see her, she snarled, "Live with it!" I gave her several options to resolve the situation and told her to call me when she was ready. She had effectively discarded me for good.

Jean called me a couple of weeks later and asked why Mother and I weren't speaking. I told her what Marlene and I had discovered and how we felt about it. I offered her the same strategies I had offered Mother to move forward, and told her to call me when she was ready to do so. I haven't heard from her since either.

By then I knew that with Dad out of the house, Mother no longer needed me to protect her from him. I continued to look after Dad, but I haven't seen or heard from Mother except for one time, and I'm fine with that. The apple doesn't fall far from the tree, apparently.

Marlene went to the house in Morden every week and knew what food Dad was eating. When we compared that against the shopping I had been doing for them, we came to the realization that he may not have been getting enough to eat. This was probably why he was weak and falling on the floor. After I moved him to the assisted living facility, he didn't have that problem again. I wondered if Mother had been deliberately depriving him of food to get him out of the house.

The last few times I went to see Dad, he wasn't in his room. When I asked the staff where he was, they told me he'd be somewhere in the hallways in his wheelchair. And that's exactly where I found him. I got the distinct impression that he was looking for a way out.

***

I DON'T REMEMBER WHY I HAD DRIVEN down to Ninette, but something told me to go there. As you drive down into the valley on Highway 18, and look across at the other side, on the top of the hill is the graveyard where my grandparents are buried. I glanced up there as I was coming down the hill, and I could have sworn that the two of them were sitting up on the hilltop looking east towards Morden. I went out to Grandfather's farm where Dad and I had hunted deer. It was a nice day, and I had lots of nostalgic thoughts about my experiences deer hunting with him there.

About three days later, I was up early in the morning. I was helping to manage the Prime Meridian Trail, and had committed to guiding some people along it. It was about six-thirty and still dark outside when the phone rang. I picked it up. It was the home in Morden.

A woman's voice on the line asked, "Is this Mr. Massey?"

I said it was, and then her voice broke and she couldn't speak.

I said, "It's about Dad, isn't it?"

She whispered, "Yes."

I told her to take all the time she needed, and waited quietly while she regained her composure. She told me she had gone into his room to get him out of bed and into his wheelchair. She had then turned to work with the other resident who shared his room. She heard a commotion and looked back to see Dad trying to get back into bed from his wheelchair. She went to help him, and he died while she was holding him in her arms.

I thanked her for what she had done and for being there for him. I said goodbye, and my first thought was how nice that he could die in the arms of a caring woman. I hoped that I would be so lucky. All the arrangements were already in place, and I made one phone call to the undertaker and one to Marlene.

I got into my truck and headed out to meet the party I was guiding on the trail that day. It was a partly cloudy morning in March, and the sun had just come up in the east. The rays of sunlight were breaking through the clouds and shining on the road ahead of me. Suddenly, Dad was there, just outside the truck, not three feet away. He would have been about eighteen, with long brown hair and a short shaggy beard, but there was no doubt it was him. He was youthful, carefree, and pure love. Up in the clouds behind him were my grandfather and grandmother, my aunt, my uncle, and a number of others. A feeling of utter peace came over me as I sensed his thoughts, "My good son, I am as one with you." And then he was gone.

I believe he came to settle unfinished business with me, and I am so grateful he did that. It changed everything.

What he couldn't do for me in life, he did in death, and that showed me the way.

CHAPTER THIRTY-ONE

The On-Again Off-Again Colony

I NEVER TALKED TO DAD ABOUT THE hog barn situation. He had volunteered in 1941 during the darkest days of the Second World War because his brother was already overseas and he felt the situation was desperate enough that he needed to do his part. He believed he served to protect the rights and freedoms of his family and fellow Canadians. I know he would have been terribly disappointed by the government of Manitoba for not protecting the freedoms and way of life of rural Manitobans like us, and it would have made him question why he and his family sacrificed so much, why he risked his life, and why others had lost theirs.

A Conditional Use Hearing was the only recourse a community had to stop a hog barn. In our case, the colony got around that by overbuilding the barn under the guise of replacing existing facilities. The present government has weakened even the possibility of a community doing something by allowing rural municipalities (RM) to forgo a Conditional Use Hearing if they choose.

The governments of recent years have put in place regulations that make it virtually impossible for the average citizen in the rural area to do much, or indeed anything, about a hog barn. They have to rely on their local RM government to deal with the problem, but often these people are not trained, nor do they have the expertise to handle these complex issues. And if you have an unscrupulous operator and a rogue municipality, possibly supported by provincial government departments, all ready to turn a blind eye to an obvious infraction, your chances for justice are pretty remote.

All the regulations favour the big corporate producer, such as the colony. When it comes to the environment, the province claims they have the toughest regulations in the country, but they are seldom enforced. It's a sham. In the case of Big Island, the rule of law that our Prime Minister is so proud of does not seem to apply.

***

THE OMBUDSMAN'S OFFICE HAD CONTACTED ME ON May 2, 2019, and indicated they were suspending their investigation while our lawyer did his work on our behalf. In July, I contacted the investigator to say we had finished that part of our process and asked for the investigation to be reopened. She said that the investigation had been suspended indefinitely, and there were no plans for reopening it. I told her that was not what I had understood from our previous discussions, and that I never received anything from her in writing. I do remember she often called when I was busy in the shop, but because she spoke softly and quickly, I obviously hadn't heard everything she said. Anyway, after that, I refused to talk with her on the phone and asked her to communicate by email.

Rather than push them to reopen the investigation, I decided to wait until Young Joe had set up the meeting with the colony. I wanted to see where that went, and was also waiting to see what I would get after requesting the Manure Management Plan (MMP) the colony needed to complete in July 2019.

Then, at the beginning of September, I got an email from the investigator announcing that she was leaving the Ombudsman's office. I took the opportunity to complain to her supervisor about the way this entire investigation had preceded.

***

AT THE END OF JULY, OUR GROUP, the Concerned Citizens of Big Island (CCBI), met for a poolside barbecue at Joe's place. Joe and his wife have a lovely setting and some people new to the district were there. It was a beautiful evening, the food was great, new friends were welcomed, and old acquaintances renewed. I told the group we had put out feelers about arbitration with the colony, and we discussed other things that we could possibly do. The smell had been better that summer, and we wondered if perhaps the colony had reduced their herd until they were in compliance. We wish!

Young Joe was finally successful in organizing a meeting with the colony in late August. He had a two-hour session with them, at the end of which he was hoping they would agree to a meeting with CCBI. From the outset, they were pretty focused on me and how I seemed to be the source of all their problems. Joe indicated that I was the spokesperson for a community group, and as such, I was putting forward the concerns for all of them. Being very competitive, I think Sam just saw this whole thing as a struggle between the two of us. Peter and the other leaders who were present did not seem interested in a meeting, but Sam was beginning to think that perhaps it might be a good idea.

Young Joe left it for a month and went back to see them. After even more discussion, they agreed to a meeting with CCBI on the evening of October 2 at the colony. In preparation for that meeting, we decided to put together a series of questions and send those over to the colony beforehand, for their consideration. I included a preamble that encapsulated the history of events up to the present, and our concerns with the colony's actions. That created quite a stir with Sam, and he called Joe and said he wanted a fresh start. Joe was alarmed, and when he called me, I asked, "Is he threatening to cancel the meeting?" Joe said no, so I suggested we have the meeting and see what happens.

Finally, October 2 rolled around. The meeting was at the metal industry building on the colony. Dorothy and I pulled up to the door, and Sam was waiting outside for us. "Hello, Billy," he said as I approached him. I asked him how he was doing. I usually try to have something lighthearted to say when I am in the company of the Hutterites at this colony. It eases the tension and they usually respond in kind. Even after all that has happened, I still feel a kinship with them.

We walked together to the boardroom of the complex. I was surprised at how extravagant it was. There were two floors in that section. A winding staircase led up to the second floor, which looked to me to be more offices. I thought this was unusual because Hutterite homes and buildings are often quite austere. It was clear they were doing very well with this business.

In all, there were seven of us from CCBI, Sam, Peter, and Walter—the third brother who was the farm boss—and a young man from the colony that no one from our group seemed to know. We agreed to use the series of questions that I had provided ahead of time to proceed with the meeting. Sam said at the beginning that they would do whatever it took to be good neighbours and keep the peace. There was a conciliatory tone on both sides and hope for a resolution that would take us beyond the impasse where we found ourselves.

Sam kept insisting that he wanted a fresh start. I said I agreed with him, and in order to have a fresh start, the first item needed to be a count of the number of animal units (AU). The colony wanted to know if it was really necessary for us to know the number of animals. I replied that it was our right to know, and I persevered in requesting an accurate count of the hog numbers on an ongoing basis.

About ten years earlier, I had received a letter from their veterinarian that indicated he had done a count and the colony was in compliance. I suggested that we would be comfortable with that veterinarian, who was known as an honest man, doing two counts per year for Muddy Woods and giving us a copy. Sam indicated that *he* was doing the counts at that time and sending the numbers to Muddy Woods—clearly, not an independent person checking for compliance as we had hoped. Talk about putting the fox in charge of the hen house.

We talked about the MMPs. The colony revealed that the lagoon had a capacity of only 5.3 million gallons when it was supposed to be 12 million. That explained to me why they had made unengineered modifications to the lagoon—they were trying to increase its capacity by building up the sides with earth and clay. I don't think it worked, and in any case, they were stopped by Conservation who told them they had not followed proper procedures, and needed to get an engineering firm to oversee the modifications.

I asked Walter how they managed if they were spreading manure only once a year. He suggested I was naïve, which I thought was an interesting term, and quite sarcastic, so I asked him to enlighten me. He told me they were spreading twice a year. Then he asked if I thought it was worse to have a little bit of manure spilling into the ditches from the fields, or if I wanted to see it running over the sides of the lagoon? I told him neither, and my thought was that if there weren't so many pigs in the barn, this wouldn't even be a problem. The young Hutterite beside me snickered and made some obviously rude remarks in support of Walter's comments. I didn't understand what he said, but I noticed that his comment amused Walter and annoyed Sam. I then asked Walter if the colony used any other commercial fertilizer, and he said they only added sulphur because hog manure did not contain that chemical. Walter smirked at Sam and said they needed more pigs to have more manure for fertilizer. Sam returned a dirty look. I think he didn't want Walter annoying us.

I have heard from Hog Watch that hog producers in Manitoba have lost money in the past eight out of ten years, while Maple Leaf and HyLife, the two large packing plants in the province, are making forty percent profits. I could see that the addition of manure as fertilizer in their farming operation would probably offset some of their losses.

When we asked about the installation of test wells around the manure storage facility, Walter said they were already in place. We had no idea they had been installed, which was not all that surprising since the colony hadn't been talking to us for 12 years. When he said the government had made them drill those wells about ten years ago, I asked if they had seen any of the test results. Walter immediately said the colony had nothing to hide, which was an interesting comment, and then said they did not know if the water had ever been tested. I offered to follow up with Manitoba Sustainable Development to determine whose responsibility it was to ensure regular testing and compliance with the guidelines. I was pretty sure it was the colony's responsibility, but I wanted to find out for sure.

The colony had gone to Muddy Woods back on March 22, 2016, and asked permission to build a feed mill. Muddy Woods had scheduled a Conditional Use Hearing and claimed they had given notice, but only to the members of our group who qualified to be informed because of their proximity to the colony. Not surprisingly, none of those members recall ever receiving any notification of the Hearing, which raised some serious concerns among all members of CCBI about the way Muddy Woods was handling this entire situation. None of this was a surprise to us, however, because we already knew that Muddy Woods had been refused entrance to the local planning district as a result of their questionable planning activities. The problem with that was without the planning district's input, only the council themselves and the RM's building inspector were in charge of this proposed construction.

Our fears were heightened when we discovered that the actual building of that mill was done by the colony itself. When I requested and received a copy of the Conditional Use Permit, I couldn't see any requirements for industry standards or environmental concerns. The feed mill appears to be finished, at least on the outside, but because it is not yet operational and we don't know what effects it will have on the community, we are very concerned.

The colony told us they had been using an additive in the manure that made the decaying process more efficient and the manure easier to handle. This accounted for the fact that the smell was not as bad last summer. So much for our hopes that they were in compliance.

I put all our cards on the table and told the colony that CCBI and its members had limited options because the current laws and regulations favoured the producer. The colony countered that they felt oppressed by the current regulations in carrying out their day-to-day operations. They also indicated they did not plan on expanding the present facility, and were trying to follow the guidelines for injecting manure into the soil. They also admitted they were aware that we knew about the manure being injected in the rain on occasions, but their attitude was what would you rather have—effluent in ditches, or running over the sides of the manure storage facility?

We ended the meeting by outlining three action items—Sam would follow up on the possibility of having the colony veterinarian conduct a count, I would contact Sustainable Development about results from the test wells, and I would create a community newsletter and review it with Sam before distribution.

When we left the meeting, people were feeling pretty positive about what had happened. One member asked me what I thought, and I said I was cautiously optimistic, but that time would tell.

As Dorothy and I drove home, we talked about what had just happened. I said, "Wouldn't it be great if the colony follows through on what we discussed tonight. What a wonderful ending to my book!" But as I got closer and closer to home, a strange feeling almost like depression came over me. I told Dorothy about it, and after considering it for a while, I realized what it was. I wasn't ready for this struggle to end, at least not this way. It just seemed too easy, and I didn't believe it *was* coming to an end, nor that we would be successful. I think deep down, I had a gut feeling of what was to come.

***

THE NEXT MORNING, I PHONED SUSTAINABLE DEVELOPMENT about the test wells. They put me in touch with the officer who looked after that department. When I explained the situation, he indicated that they didn't know about the manure storage facility at that location, never mind any test wells. Good old Conservation had done it again—told the colony to comply with the regulations, and never followed up, apparently losing track of this facility altogether. I asked him who was responsible for sending in the test samples, and as I suspected, it was the colony. He assured me that Sustainable Development would be looking into this and the colony would be required to send in samples once a year. You have to wonder about that, but that's the process.

When I asked if the community could request the results of the testing, he indicated that was private information covered under the Freedom of Information and Protection of Privacy Act regulations. We could get a copy of the results, but all significant information would be redacted. So I asked what would happen if there was a concern uncovered by the testing, and he responded that Sustainable Development would then review the colony's business plan with them. And that was it. It wasn't clear if that meant the facility would need to be repaired or replaced, or what would happen, and we still don't know. I sent all of this information to our group and to Sam at the colony.

I waited about a week before contacting Sam to find out if he'd had a chance to speak to the veterinarian about conducting a count. When he responded, "Animal units were decided fifteen years ago, and we are moving forward," my heart sank. I knew at that moment the good will we had seen at the meeting was gone. Even I was surprised at this complete change in direction so quickly.

In my email response, I reminded Sam that the colony had agreed at the meeting that a count of the AUs would take place and they would approach their veterinarian to do so. If they were not prepared to keep their word in this matter, then there could be no movement forward. If they did not want a count, then we could only assume that they were not in compliance. I suggested they consider carefully before going this route, unless they wanted to see a continuation of the situation as it was before. I asked if they would meet with us the next day.

His response was swift and bold, literally. "It was talked about but not agreed, it was a suggestion. I said it could be done, not that it will be done." The last sentence in bold red said, "I will decide when the pigs will be counted. Not Bill."

In an effort to be friendly and keep the door open to continuing negotiations, I thanked him for his prompt response and reiterated the action plans we had agreed to. I suggested we should take this matter one step at a time, and said that yes, that it was his decision when the pigs would be counted. There wasn't any question of that as long as it was done in a timely manner. I even invited him to our place for tea to talk it over and told him to bring the others. If that was inconvenient, I said I could come to the colony. I also sent him a copy of the extensive notes one of our group had taken at the meeting on October 2, and asked him to advise us of any errors or omissions. No surprise there was no response.

After about five days, I emailed him and said that judging by his lack of response, I was going to assume he was not interested in continuing negotiations. I invited him to let us know if he changes his position on that matter. I said a count was necessary to be able to move forward and hopefully rebuild the relationship between the community and the colony. I am still waiting for a response.

I called a meeting of CCBI on October 24 to discuss our next move. We wanted another legal opinion and some us were trying to find a lawyer experienced in municipal law that we could talk to. We decided, for the time being, to write a registered letter to the colony with the following points:

1. We would need a count if the odour was bad, which would be determined by the smell charts kept by the neighbours.

2. When we requested a count, we would want that done within a week by the veterinarian who would then share the results directly with us.

3. We wanted to see the results of the test well samples.

4. We would need a commitment to be sure that any spreading of manure was done so that all guidelines and rules were being followed.

5. We wanted a meeting with all three of the RMs affected, the colony, and ourselves to get dust control and traffic issues resolved satisfactorily for everyone.

6. We wanted to meet at least yearly to discuss any issues that might arise.

We concluded by expressing our wish to resume negotiations.

A few of us worked on the letter, emailed it around to others for comments, and a number of people offered their input. The letter was sent to the colony on November 10, and they were given till the 20th to respond. We decided to again ask the Ombudsman to resume their investigation.

November 20 came and went without any response from the colony. I had been communicating with the Ombudsman's office, and we had been assigned a new investigator. The investigator was interested to know if our attempts to reopen negotiations with the colony had been successful. I waited till November 26 to email the Ombudsman that our attempts had not succeeded, and then I requested the resumption of the investigation. I sent Sam a draft of a newsletter that I had prepared for the community, and gave him three days to respond with his comments. Not surprisingly, I heard nothing.

On November 29, I sent a newsletter out to the community talking about our meeting with the colony. We told them about the test wells, the enzymes added to the manure to mitigate the smell, and that there had been no modifications made to the barn since it was built as stated by the colony. We told them that when we pressed the colony to find a mutually acceptable third party to do the counts, they abruptly ended communications with us. We had noticed the intensity of the smell from the barn had increased since our meeting with the colony, and

I suggested that if the smell worsened again, we would consider a complaint to the Manitoba Farm Industry Board.

I sent another smell chart to the community and asked them to keep track of the smell until the end of March. I also sent copies of the community newsletter to the council at Muddy Woods and our MLA. He responded that he was hopeful we could come to some kind of a compromise going forward. I told him I would be in touch once the Ombudsman had completed their investigation. We did not hear anything at all from the council.

The Ombudsman indicated that they hoped to restart their investigation first thing in the new year. By early February, the investigator had been in touch with the Department of Agriculture, but is yet to meet with us.

***

I DECIDED AFTER CHRISTMAS THAT IT WAS a great time to get started on this book. We were pretty much unable to do anything else until the Ombudsman's investigation was completed, and while I waited, I thought that putting together all the information about my life and this ongoing situation with the hog barn would be a good use of my time. It's amazing how writing down one's stories provides clarity and renewed enthusiasm for the road ahead. And since I didn't know what was on the road ahead, I told myself that writing would help.

On February 22, I met the new NDP Agriculture and Environment Critics at the provincial council meeting. I asked them to come to Big Island so I could show them around and we could discuss the hog barn at the colony. I also invited a Hog Watch activist from Steinbach so as to give them an overview of the provincial situation. One of the critics, his aide and the activist came, as well as a member of CCBI, and we all climbed into the activist's SUV to tour the area. We drove around the section where the barn was located and out to the historic Grant's Lake Wildlife Management Area. As I talked about the situation, they listened attentively and asked a number of questions. Afterwards, we all sat around our kitchen table, drank tea and coffee, and ate a dessert that Dorothy had prepared. Our conversation about hog barns was very productive.

I had asked the activist to talk about the situation in the province generally, and she did a great job. Of most interest to me was what was happening in her area and elsewhere in the province. I knew that if you build a barn of less than 300 AUs, you can avoid the Conditional Use Hearing. That way you do not have to answer to the community. In the Interlake, there is a situation where four smaller barns are clustered around a common manure storage facility, each on a different quarter of land owned by different members of the same family. These barns combined house almost 1,200 AUs with no possibility for community input.

East of Steinbach, swamps are being bulldozed and barns built on them, and the manure is being broadcast on the surface because it cannot be incorporated. As a result, the nutrients are being carried into watercourses, which is, of course, an ecological disaster, not just because the manure was being deposited in riparian areas, but also in the swamps themselves. These great carbon sinks are being destroyed, which will obviously have disastrous effects on the environment.

In addition to the provincial government making it possible for RMs to forgo the Conditional Use Hearing if they choose to, each RM can also determine how many people are required to show up in order for the hearing to be legitimate. And more often than not, it's difficult to get the required number of people to attend a meeting at the same time, given how spread out we are and the type of work being done on each farm. Such requirements show a blatant disregard on the part of the province for community concerns about these operations, and are just another way to make it difficult for us to oppose expansions and other potentially negative community impacts.

One day in March, I went to get the mail, and when I got out of the car, I heard a voice saying, "Mr. Massey, how are you?" I looked around and realized it was Young Joe, who I hadn't seen for a while. He said,

"I should tell you about the meeting I had with Sam the other day." Of course I was all ears. He explained, "Yeah, Sam called me up and told me he'd like to meet with me privately."

Joe was heading into the colony to meet with Sam when he met Peter driving out. Peter turned around and followed him. Young Joe was annoyed, but couldn't do much about it; he knew it meant he wouldn't be meeting with Sam alone.

Sam began by referring to our previous request for a veterinarian to do a count of the pigs and to let us know within a week. He said something about the fact that he couldn't organize a count within a week, which didn't make much sense to Joe because we hadn't asked for the count to be organized within a week. All we had said was that when we *did* request a count, we would want that done within a week.

Then Sam told Joe the main reason for the meeting—he wanted to let us know he had retired from managing the hog barn as of the first of the year 2020. I smiled and said to Joe, "Ah, so I *did* outlast him."

Sam also said to Joe that the colony had won, and all they had to do now was wait for me to move away or die. When I heard that, I wondered briefly if the new manager might be less competitive than his predecessor, and then I laughed heartily at what Sam had said. I told Joe, "Well, I'm still here and I'm not dead yet!"

***

FEBRUARY 21, 2020, THE SMELL IN MY yard from the barn was as bad as I've ever experienced. I called the reeve of Muddy Woods and invited her down to share the experience. I took the opportunity to tell her they had too many pigs in that barn, and the RM of Muddy Woods needed to enforce The Planning Act. She said that they were in compliance, and there was the matter of having trust in the colony. That was the wrong thing to say.

After a stern lecture, I asked if she was coming to experience the smell for herself. She said she was, so I left on the lights and waited and waited. The next morning, when the smell was nearly as bad, I called her again. She told me she'd gone over, picked up Sam from the colony, and brought him over this way, but he couldn't explain why the smell was so bad. I told her I could explain it, and it was simply because they had too many pigs in the barn. She then told me that the Department of Agriculture was pressuring them to allow the colony to fill the barn. I said we'd also talked to the Department, and they had told us that they would never tell an RM to break their own bylaws.

When I said that we believed the council was discriminating against our group, she complained that the RM had spent seven thousand dollars on legal fees because of us. I told her the money would have been better spent on hiring a mediation firm to help us all sit down together—the colony, the RM, the department, and ourselves—to resolve the problem. When she said she had only one vote on council and couldn't do much, I countered, "As the reeve, you have a lot of influence and you should work diligently to address this problem."

I told her the only reason we were inactive right now was because we were waiting on the Ombudsman's report, but once that was completed, we would be back and moving forward in whatever way we could. There was an audible groan on the other end of the line.

CHAPTER THIRTY-TWO

Marlene

WHEN MARLENE FINISHED HIGH SCHOOL, SHE ENROLLED in a Bible school in Saskatchewan. One of the members of the Pentecostal Church in Glenella had sponsored her. It was a three-year program, with summers spent doing missionary work. It did not go that well for her; she was not cut out for it, and two months before she would have graduated from the program, she dropped out.

My parents were quite concerned because they thought that her sponsor would want his money back. But he was a magnanimous man and said repayment was not necessary. I think that was all my parents were worried about; they were blind to the fact that my sister was in crisis.

Marlene came home for a little while until she decided what to do next. She took a secretarial course at Assiniboine College in Brandon and soon got a job in that city. She had trouble living and being on her own, but when she moved into an apartment, she was quite frightened and felt threatened by some of the others there, especially the young men. I was teaching in a little community outside of Brandon, and kept in close contact with her. She was having a lot of trouble adjusting to her new life, and really appreciated my support.

It definitely helped when she found a church in Brandon that she liked and started teaching Sunday school there. She met a dairy farmer on a Sunday school outing to his farm, and she ended up marrying him. He was a little older and was very good to her, being patient and supportive of her all her life. They were happy together. Eventually, she quit her job and started working with him fulltime on the dairy farm.

When she was about thirty, about five years after she got married, she got quite sick and extremely tired. She had what I think was a psychotic break. I got a call from my mother, who had been summoned there by her husband. Marlene had become quite irrational, and they had dragged her kicking and screaming into a psychiatrist's office. He took one look at her and said there was nothing he could do; he suggested they bring her back when she was calm.

Mother called me after she and Marlene's husband got her back home. I listened to the story of what had happened and later talked to the school psychologist, asking for some advice on how I should respond to this situation. She gave me some good ideas, and I went out there one Saturday morning with my first wife, who understood Marlene and was supportive of her. She was quite sympathetic, especially when she realized how our parents were treating her.

We got there about ten o'clock in the morning and knocked on the door. When Marlene greeted me, she started apologizing over and over. Her eyes were black and hollow, and she had obviously not been sleeping. She was saying irrational things, like my wife was going to choke her by shoving a glass down her throat. I embraced her and gently eased her over to the couch. I sat down beside her with my arms around her.

She apologized for things that had happened in our childhood. She apologized for chasing me under a barbed wire fence when my back got scratched. She even apologized for running back to the house to get the hairbrush that time when Dad got mad. I kept saying to her, "It's alright, Marlene. It's not your fault, Marlene. You did nothing wrong, Marlene."

I remember telling my mother to stop trying to make Marlene feel guilty and blaming her for what had happened; I knew that was going on, too, and likely making things worse. After six long and intense hours, Marlene finally calmed down. I took her to her bedroom and put her to bed. I was emotionally drained at the end of it and really concerned. She slept the clock around. When she got up the next morning, she was her old self again. I alone had brought her back. After that, she was always careful to get enough sleep.

Her attitude and behaviour changed towards me compared to what it had been before, and for a while she was calmer, more relaxed, and appreciative of both the support I was giving her and my concern for her. I made a point to stay in closer contact with her on a monthly basis, but after several years, she started withdrawing again, and I didn't see much of her for some years. She did, however, consistently have something to do with my family, especially during the holidays. Marlene and her husband would always bring presents for the children at Christmas and attend their birthday parties. She did her best to stay connected with them, and my first wife appreciated her because of how she cared for the children.

Things changed again when there was a crisis in my family and my marriage broke up, but this time Marlene was there to support me. We reconnected and got progressively closer again as time went on. As our parents aged and began to decline, we had more and more to do with each other, working together to care for them.

I remember my mother asking me one time what had happened to Marlene. I had my answer ready. I told my mother it was the childhood we had experienced, and that she was still dealing with issues from that time in her life. Not surprisingly my mother never asked for my opinion on the subject again.

Mother was still being nasty to Marlene. After Mother had discarded me, I asked Marlene why she was still going there and looking after her. She and her husband were cleaning the house and looking after the yard. That was when I finally told her what I had overheard our parents say about us that night when we were children. She was not surprised; she may possibly have heard it herself, although she didn't say so. Her response to me was, "I am the oldest, and I feel it is my responsibility to look after her."

I wasn't surprised that a sense of duty kept Marlene coming out to help our parents. I greatly admired her for that, but I knew it was not easy for her. Later on, when I talked to her husband, he told me that when they drove back to Brandon after spending some time in Morden with my mother, Marlene would shake most of the way home.

Following the incident with the cheques and after Mother ended her relationship with me, Marlene and I teamed up in dealing with our parents. She became even closer with me and Dorothy and our family and came every year for Christmas. One year, she said it wasn't Christmas until she had been with her own. Indeed, our children were like her children, and they loved her dearly.

***

JUNE 2008, I RECEIVED A LETTER FROM Marlene about her upcoming surgery; there was a mass in her abdomen that needed to be removed. She seemed worried about having the procedure, and both Dorothy and I thought the letter was her way of saying goodbye. I believe she knew what was in store for her.

The operation took place on a Wednesday in early August. I called her husband the next day, and he told me she was in a lot of pain. On the Saturday, we drove out to Brandon to visit her in the hospital. She hated being there; she was such a private person and a hospital was just not the place for her. The pain wasn't as bad that day, and I sat on the edge of the bed with my arm around her. I asked if there was anything I could do for her. It was the long weekend, and I got the impression that she wasn't getting the care she needed. They had holiday staff on at the hospital, and neither the nurse nor the doctor seemed to have enough time to really talk with Marlene about her pain. When I offered to talk to her physician about it, there was no uncertainty in her voice when she told me she did not want me to intervene and that she wanted me to go home.

I told her I loved her, she told me she loved me, and I did as she asked. I believe she knew what was coming and wanted it that way. She did not want to be saved again. On Monday, she signed herself out of the hospital. Thursday evening, she collapsed at home. Her husband called the ambulance, and she was taken to Brandon General. She died on the operating table in the early morning of Friday, August 8, 2008, from septic shock. She was sixty years old.

Her husband woke us up with a call at two in the morning to tell us what happened. My first thought when I heard the news and shared it with Dorothy was that she did not have to suffer anymore. Dorothy put her arms around me and comforted me. We didn't get much sleep after that.

The next morning, we drove to Brandon to help her husband with the arrangements. He asked me to do the eulogy. He told me he wanted me to do it because he felt I was the only one in her family who loved her.

When it was all done and everything had been arranged, Dorothy and I came home. I called my son with the news. He was serving in the Canadian Navy on Vancouver Island at the time, and made arrangements to fly home for the funeral. I called my daughter and told her what had happened. I also called my first wife and told her the story. She was shocked that things had gone this way and wanted to come to the funeral with us. Her support, especially for the kids, was much appreciated. My daughter-in-law called from the west coast out of concern for me and talked to me at length. Another brother-in-law was very supportive and offered to drive all of us in my van on that day. My cousin even came out from Victoria.

The day of the funeral, we all assembled here on the farm before heading to Brandon. When we got to the church, there were a lot of people there, and during the eulogy, I made a point of thanking them all for coming to say goodbye to my sister. I also shared some of the wonderful things about her that made her special and so very important to me—

> She cared deeply for her parents and, with her husband's help, looked after them faithfully in their declining years.
>
> She maintained a close bond with me and my family, and took a special interest in her niece and nephew and their partners.
>
> She had a sweet, gentle spirit, and a kindly heart, and her generosity knew no bounds to all who knew and loved her well.

Then I shared these words, slightly modified from a favourite song that Aaron Neville recorded, "I Bid You Goodnight"—

> Goodnight, dear sister, sleep and take your rest.
> Lay down your head upon your Saviour's breast.
> We love you well, but Jesus loves you best.
> Goodnight, goodnight, goodnight.

I said those words with the deepest emotion and then sat down beside Dorothy. It was the most difficult thing I ever had to do in my life.

When we gathered at the graveside, Dorothy and my ex-wife stood together to one side. Jean, her family, Mother and Marlene's husband were on the other. People gathered behind us. I stood alone near the edge of the grave as the coffin was lowered into the ground. My son and daughter came and stood right beside me, her head on my shoulder, his arm around me. We stood that way until the minister from the Alliance Church had performed his duties. We three wept; I don't know about the others.

Staying by the grave for a little while, I comforted my children. I felt very much supported, and have never since felt as completely alone as when Marlene died. I think of her every day.

CHAPTER THIRTY-THREE

The Journey Continues

WE DID NOT SEE OUR FOUR GRANDCHILDREN for two months this spring because of the pandemic, and we missed them terribly. On June 21, 2020—Father's Day—my son's family with their two children came out to the farm for a socially distant picnic barbeque. I was so glad they were there and I spent most of my time with the kids, either playing with the new lambs or climbing on the bales, all the while making sure they didn't hurt themselves.

My granddaughter is turning out to be such a lovely girl, and I'm so proud of her. When she was younger and they stayed over, she would get up early with me to watch the day being born. She would snuggle up on my lap in the big easy chair and ask for stories of Princess Ariel coming to the farm. Now she is a young lady, she doesn't want to do that anymore, but she still likes spending time with me. I've taken her for horseback riding lessons and we went on a trail ride this summer.

When my son-in-law had surgery, I went to their place to mow the grass. My grandson, their oldest boy who's six, ran up to the other side of the fence and told me he loved me. We talked about social distancing to him because of COVID-19, but he soon forgot and came running towards me again to give me a hug. My daughter intervened, scooped him up in her arms, and told him she would give him a big hug from

Grandpa. I was so, so pleased with him about wanting to hug me! When he comes here to the farm, he doesn't want to leave.

My daughter's other son, my youngest grandson is confident, happy-go-lucky, and unstoppable, just the way you would expect a three-year-old to be. The other day when I called his dad, they were all in the truck, and I also got to talk to my daughter and the older boy. When I tried to speak to the youngest one, his dad told me he had put his head down. At first, they thought he was being shy, but because it went on for some time, they realized he was missing his grandpa. I am so lucky. He and his brother came to stay with us in the middle of August for a few days. He loves to help, and had his little red wagon in the garden helping Mama Dot gather carrots. I have hopes for him to help me on the farm when he gets older.

My son's boy, my oldest grandson, loves to explore our farm. We go together to the creek that runs along the edge of the property. We make our way through the swamp grasses that have been pushed down by the snows of winter and now tangle around our feet. He tells me the swamp monster is trying to get us, and when he sees a gnarled old stump, he excitedly calls, "Look, Grandpa, he's poking his head up to look at us!" That was exactly what I would do when I was seven years old. I'm grateful that I can offer him these experiences that Dorothy and I were so lucky to have when we were children.

The adults have their toys here as well—go carts, dirt bikes, snowmobiles—and they come to play with them frequently. My son and I go for walks together when he is here. Those are times when we can share our inner feelings with one another. Sometimes my daughter and I walk and talk, too, but usually when they come for a visit, they leave the boys with Dorothy and me so that she and her husband walk around the farm by themselves.

***

IN THE SPRING OF 2018 WE HAD a drought. When my oldest grandson and I walked out to a slough a quarter mile north of us, he asked why the wind was stinging his face so much. I explained to him that the air was filled with dirt, and if he thought it was bad for him, to imagine the impact of that blowing soil on young crops, the tops of which were being cut off. I added that some of the seeds were even being torn out of the ground.

He looked at me and said, "That's not good, is it, Grandpa?"

I told him, "No. No, it isn't good."

Personally, I think we could plant more trees and native prairie grasses, grow more food for ourselves, and walk more, drive less, but that's just my opinion.

My family loves this place, but we have to check what the wind direction will be before we plan any outdoor activities. Sometimes we will cancel or spend the time indoors because of the strong smell from the pig barn. Trees will mitigate the smell from the barns, but many bluffs, road allowances, and woods have been eliminated to accommodate the large farming operations we see today.

***

IN SOME AREAS IN THE SOUTHEAST OF the province, you can stand at any given point and count thirty or more hog barns. They all use liquid manure systems, and the manure is incorporated into the ground. A cultivator drags a hose filled with effluent that is pumped from the manure storage facility. The range that the systems have is determined by the length of hose that the cultivator and tractor can pull; usually it's not more than three miles. This limits the range of the spreading and means that those fields close to the hog barn are saturated with nitrogen and phosphorus.

In 2007 the provincial government addressed this problem somewhat by placing a moratorium on the building of hog barns in the Red River watershed. The current government has removed that moratorium. In her 2017 article, Alexis Stockford indicates that the levels of phosphorus being transported off the fields by rain and surface water is too high and not sustainable.[4]

The small farm is disappearing. You don't find many small mixed farms anymore where pigs are raised in straw-based manure management systems known as biotechs. That was the way pigs were raised in the past, how lots of smaller operations raise pigs for food now, and how Dorothy and I raise pigs today. At one time, small farms like ours were scattered all over the province and the manure was spread as a solid waste on the fields. Because farmers were able to spread the manure on *all* their land, it avoided the issue of huge concentrations of nitrogen and phosphorus that occur in factory farms today.

I believe the only reason that we don't have even more hog barns is that the producers have been facing many challenges. Shannon VanRaes outlines some of their difficulties—too numerous to mention here—in her article, "Where have all the farmers gone?" posted June 16, 2016, in the Ag Canada newsletter. In the January 14, 2018, issue of the *Virden Empire-Advance*, Larry Powell stated Maple Leaf and HyLife, the two large packing plants in Manitoba, have received interest-free loans from the federal government that they may not have to repay. Maple Leaf, owned by the McCain family, is renting pigs to farmers to encourage production. HyLife, a Japanese and Korean company, is buying land and building barns in southern Manitoba, hiring people to manage them.

According to Statistics Canada, the average age of Canadian farmers has reached 55 after rising for decades, and 92% of farms have no written plan for who will take over when the operator retires. It also found there were more farmers over the age of 70 than under 35. HyLife is a multinational corporation for which farmland has become a good investment opportunity in Manitoba. Some have suggested this is a new form of colonization, which would be ironic given that we took the land from the Indigenous people and exploited it, and now it might

4 Alexis Stockford, "The Phosphorus Conundrum: Low soil levels meet Lake Winnipeg pressure," *Manitoba Co-operator*, March 6, 2017, https://www.manitobacooperator.ca/news-opinion/news/local/the-phosphorus-conundrum-low-soil-levels-meet-lake-winnipeg-pressures.

be our turn. All I know is it's our responsibility to protect and nurture this land for future generations.

I believe that many farmers are caught in the need to expand to survive. Fuel, fertilizer, seed, and machinery costs are astronomical. Investors, some using foreign money, are driving up the price of land, making it very difficult for young Canadian farmers to get started. Several years ago when I asked a young farmer who was a neighbour of mine how things were going, he said that the bank was still letting him farm. After the drought in 2018, I see he has taken a job and rented out his land. That speaks volumes.

There are rural municipalities (RM) in Manitoba that will not allow hog barns to be built within their borders. In Bill 19, The Planning Amendment Act, the present government had set up a Municipal Board, a group of people appointed by the province with the power to overturn the decisions of the elected officials in RMs. This is an assault on democracy. It means RMs may not be able to protect their residents and the environment from producers who are willing to challenge their decisions by going to the Municipal Board. I am amazed at the lengths this government and the previous one have been willing to go in order to support an industry that I believe in the long-term is not sustainable.

For a while, many of us were hopeful that adding amendments to Bill 19 would not only go a long way to improving efficiency in planning, but also to providing a fair and equitable voice for residents in rural Manitoba. However, that hope was dashed a few months ago after a ruling in Lilyfield.

A commercial interest had been applying to the RM of Rosser for thirteen years to develop a quarry at Lilyfield, just a few miles south-east of Big Island. That quarry had been rigorously opposed by local residents each time different proposals were presented. Residents were steadfastly supported in their opposition by the RM that denied every application that came forward.

It was a stressful fight that saw some residents report to the RCMP threats they had received from parties associated with the proposals.

When the matter was finally heard in July 2020 by the Municipal Board, many residents were not even permitted to speak at the hearing. The Board found in the applicant's favour and overturned the decision that had been made by the elected representatives of the RM's council. It made my blood boil when I heard that story from a resident, who was a former student of mine, and I offered my help.

In addition to everything else I've learned during the past fifteen years, the Lilyfield ruling convinced me that Bill 19 does not serve the best interests of rural municipalities, and needs to be repealed. We simply cannot allow democracy in this province to continue to be compromised. The enforcement of The Planning Act should be a provincial responsibility. We need to stop raising pigs in close confinement. Pigs in that situation are fed large doses of antibiotics that find their way into the environment. Liquid-based manure systems that are used on factory farms are creating dangerous levels of nutrients in the soil. The ethics of raising pigs in this way is yet another issue. I raise my pigs outside in a straw-based manure management system with no antibiotics, and I never have health issues with my animals. We have to find alternatives to raising pigs in liquid-based manure systems in confined areas, alternatives that are environmentally, ethically, and financially sustainable.

***

SEPTEMBER 14, 2020, BROUGHT A DAY THAT was windy and cold. Jim, our butcherer for the past ten years or so, was coming with his uncle Jim to butcher our five pigs. Because I had been through this process scores of times, I prepared the pen and equipment for them. I used to butcher our pigs myself, but now I get Jim to do it. I also buy our pigs from him, and give him an extra $20, telling him it's for his gas. This way Jim saves his best pigs for me.

They arrive at about ten in the morning, and the first thing he does is go over to the pen and admire the pigs. "Boy, these are nice pigs," he says. "And are they ever quiet. Much quieter than when I brought them here." I tell him we look after our pigs and they are well treated.

People know how fond I am of my pigs and I brag about them constantly. They wonder how I can have them killed, and I understand why they feel that way. I suppose it's been a lifetime of raising animals for food that makes me see this as just another job that I do on my farm to feed my family. It does, however, hit me the next day when I go to the barn and it is so very quiet. I'm reminded of the term 'a deathly silence,' and it usually takes me several weeks to get over the feeling that something's missing. But every year I look forward to getting the young pigs in the spring and doing it over again.

***

SEPTEMBER 26, 2020, WAS A BEAUTIFUL FALL day with gentle breezes from the southwest and a clearing sky. The wind was not directly from the colony's barn so we were getting only the occasional odour. It had rained a little that morning and the fresh scent of rain was mixed with the smell of the pig operation. I was working in the shop with the door open when Dorothy called from the porch, "It's our son and he has news."

When she handed me the phone, I said, "Good morning," and asked him what was up.

He answered, "I have bad news, Dad. Your mother died last night."

Jean's oldest daughter, her middle child who has made some effort to keep in touch with us, had called and let him know. I asked if he knew if Mother had died from COVID-19. He didn't know, but said he didn't ask too many questions because the girl was obviously upset. She told him there would be no service and Mother would be buried beside Dad at Kelwood. We talked for a while and I ended the conversation by saying she had never been a happy person, and that I hoped she would have a better go of it the next time around.

Dorothy and I sat down over a cup of tea and talked. She wondered if Mother had come to visit me after she died, the same way my dad had done, but I assured her that didn't happen. Dorothy and I had had a wonderful evening together the night before and I'd slept better than I usually do. I had no inclination anything was happening and the news came as a complete surprise. We speculated whether my sister would call, but I haven't heard from her. The apple doesn't fall far from the tree in her case either.

My sister-in-law and her partner phoned around five that evening and suggested they come over for a campfire and hot dogs. I talked a lot about Mother—they were aware of the story—and I filled in details of the events that had led to our estrangement fifteen years ago. They were both very supportive and I appreciated the opportunity to talk.

It was an interesting day. I tried to understand what I was feeling. I haven't shed a tear nor am I likely to. I experienced a wide range of emotions, including regret, anger, frustration, and relief. I regret that I did not have an opportunity to deal with Mother's discounting of me and Marlene while they were still alive. I was angry and frustrated about the way Mother and Jean had treated Marlene, and their steadfast refusal to give her any respect and dignity. I felt relief for Mother and myself that it was finally over.

Fifteen years ago, Mother expressed to me her frustration about getting older. I was sorry for her that she lived so see her 96th birthday, which was much longer than she wanted to. I have been very frank about her in this book, and I was a little concerned about publishing it while she was still alive. There is still a part of me that doesn't want to hurt her. One of my friends gently suggested that Mother had done me a favour and her passing was the best thing for both of us.

My son had spent some time with his grandparents when he was a young man. He had met some of the young people in Morden and they had asked him who his grandmother was. When he told them, they said, "Oh, yes, we know her, she's the bitch of 9th Street!" I felt sorry for the people who looked after her in her final years. My dear 106-year-old mother-in-law suggested I write a poem about her, and I think I will, but not today; that's something I'm going to have to really think about for a while. I know I grieved for several days, but I came to realize that I wasn't grieving for what I'd lost, only for what I'd never had.

***

WHEN I FOLLOWED UP WITH THE OMBUDSMAN'S office about the status of their report, they told me that a copy had been put in the mail and I should have already received it. I instantly felt both elation and trepidation. The investigation that had started in December 2018, and halted while there was a change in investigators, was finally finished. It had taken almost two years, during which we were unable to proceed with any other initiatives.

On October 1, Dorothy and I were coming back with supplies from Stonewall when we stopped by the mailboxes in Big Island to pick up the mail, including the report. Back at home as we sat and had lunch, Dorothy showed me some quotes by Winston Churchill sent to her by a friend. One of the quotes, I thought, was apropos for this situation—"Success consists of going forward from failure to failure without loss of enthusiasm." Because I was busy hauling round bales all afternoon, I didn't get to read the report until just before supper.

Of the four times we have been involved with the Ombudsman's office in the past fifteen years, I believe it was only twice that the investigator did a good job and supported our complaint. As much as I wanted to, I didn't have my hopes up about this time, and I was right not to. I'm sure the pandemic didn't make it any easier by not allowing us the opportunity for a face-to-face interview with the investigator. As a result, we weren't able to put a human face on the situation, our concerns, and the long journey we'd been on.

Our first complaint concerning the number of animal units the colony actually had when the process began back in 2004 was dismissed out of hand. The rationale was that too long a time lapse had occurred and the Ombudsman simply refused to go back that far in their investigation. This was in spite of the fact that we had put all the information regarding the situation at their disposal.

Our second complaint was about the behaviour of the reeve of Muddy Woods. We were concerned that she had made her decision and announced it in the local paper before we even got a chance to talk to the RM. The Ombudsman contended we'd had an opportunity to discuss the situation at council a few days after the article had been written, which was completely beside the point we were making. Nothing was mentioned about the fact that the reeve had clearly broken the code of ethical practices set out by the Ombudsman's office itself, which was *exactly* the point we were making.

I had talked to the investigator about the two situations when the council and the reeve told us the Department of Agriculture was pressuring them to allow the colony to fill the barn with pigs. If this was true, the Department was encouraging Muddy Woods to break their own bylaws. Nothing was mentioned about that in the report either.

Nor did the report address any of our specific concerns or issues, and other things we had asked about were not even mentioned. The investigator did go into a long dissertation about the purposes and uses of Manure Management Plans, and had obviously learned a lot in the course of her investigation, but nothing in the report was new to us.

After reading the report, I sent an e-mail to the more than twenty members of our group, the Concerned Citizens of Big Island (CCBI). I asked if they would prefer a socially distanced meeting or to communicate via email, and I also proposed that we go ahead and set up a meeting with our MLA, Ralph Eichler, as well as Manitoba Conservation and Climate. I then mailed a copy of the Ombudsman's report to our MLA, along with a reminder that he had agreed to meet with us after the report was completed.

That night, a member of CCBI told me that lately, the smell from the barn had been unacceptable in the village almost daily. Other members emailed their concerns about getting together given the rising COVID-19 numbers, and we decided to continue discussing the situation by phone and e-mail. We also agreed that we should meet in person with both our MLA and Conservation and Climate, and I've set that up for January 2021.

So, there you have it. We're back on the road again.

***

DOROTHY AND I DIDN'T RECEIVE AN INVITATION to Mother's burial, nor did my daughter. My daughter wouldn't have gone anyway because of her concerns about COVID-19, and it was fine with me either way. Dorothy said she wouldn't go where she wasn't welcome. And after my son thought about it for a while, he also decided not to attend out of support for me. I offered to pick up my daughter's two boys that morning so they could spend the day on the farm with us; their parents came for a lovely supper and took them home afterwards.

About two weeks after Mother died, the day dawned clear and bright, and the overnight frost lightly covered everything. I suggested to Dorothy we take a drive up to Kelwood and she was game to go. I wanted to see the grave and I was curious about what Jean had done. I had to run a neighbour into Stonewall that morning to pick up a vehicle, and Dorothy took the opportunity to call her mother and pack us a picnic lunch.

When I got back, she was still on the phone. My mother-in-law asked why I was visiting Mother now when I didn't when she was still alive. It was a great question and I didn't have a ready answer. Perhaps I hoped the experience I was about to have would help me understand that very thing. My mother-in-law often says things that are very wise and profound.

By the time we headed out, the sky had clouded over and the day had become damp, bleak, and dreary, but that didn't matter to me. I love spending time with Dorothy and really enjoy travelling with her. If one of us isn't sleeping, we talk and talk. It was early October, the leaves were beautiful, and geese were flying noisily overhead. The rain only started to fall when we reached Portage.

We pulled off #16 highway at the old gypsum loading facility at West Roc. We enjoyed some lovely farmer's sausage sandwiches and a piece of Dorothy's delicious homemade apple pie. Dorothy wanted to see where I had careened the '55 Pontiac into the Grassy River so we went that way. Then we took the same correction line road I had driven the old John Deere D at two in the morning those many years ago.

On the way I told Dorothy about my recent conversation with a cousin on my mother's side. His mother and my mother were sisters. He had called me when he heard the news, and was very supportive of me. I went to his mother's funeral a few years ago, and met an uncle I had never seen before. My cousin had spent some time with that side of the family and said there were some pretty wild characters in the mix. Mother had often recounted a favourite childhood story about her school outings to play baseball on sports days at neighbouring schools. The students would arrive in the back of an old grain truck, singing their school song:

> We're the Tiger Hills Savages, rough and tough
> We are made of the good old stuff.

Mother worshipped her father and saw herself as being like him. The word 'savage' pretty much describes them both.

***

THE GRAVEYARD AT KELWOOD IS ON THE first rise of the escarpment that is known as the Riding Mountains. It had been a beach of the glacial Lake Agassiz that had covered much of Manitoba. The mountains were in their full magnificent fall colours. We parked at the gate of the cemetery and walked in. I stopped at Mr. and Mrs. Carter's grave and said hello, as is my custom.

We then walked over to where Dad was buried. There was the fresh grave, although the tombstone was missing. I suppose it had been taken away to have Mother's name added. I could see that Jean had taken care of things. Dorothy and I sat on a retaining wall near the grave. I noticed the beautiful yellow leaves on the trees at the edge of the cemetery; yellow was Mother's favourite colour. I felt very much at peace. It was over.

We left the graveyard and went to a small restaurant at the edge of town for my tea and Dorothy's coffee. The sky cleared off and we basked in the warmth of the sun streaming through the big window beside us. The sky was bright blue with a few wispy clouds drifting over the splendour of the mountains.

***

BE WELL, DEAR READERS, AND STAY TUNED. I'm sure there will be lots more to tell.

Bill ploughing under clover as green manure with his 1950's tractor and 1920's one-way.

AUTHOR'S NOTE

I HAVE CHANGED THE NAMES OF SOME of the people in this story in order to protect their privacy.

Throughout this book, I have done my best to represent the facts as I remember them, and have used the pages of data I collected over fifteen years about the hog barn conflict for reference. Ironically, it was the delay in the Ombudsman's report that gave me the time I needed to finish this book.

It has been a remarkable and exciting journey for me to write *Of Pork & Potatoes*. I have learned a great deal about myself through the writing process and it has brought back many memories. I am rejuvenated and excited about continuing the hog barn battle. I have enjoyed interacting with and learning from friends new and old who have helped me along the way with this project. It's been a lot of work and I hope you enjoy reading the book as much as I enjoyed writing it.

— Bill Massey

GLOSSARY OF TERMS

Animal Unit – AU

A measure of the number of animals required to produce 73 kilograms of nitrogen annually. For example, a sow and all her offspring at market weight is 1.25 AUs.

Bill 19, The Planning Amendment Act (Improving Efficiency in Planning)

A Bill that introduced a number of changes to The Planning Act intended on streamlining regulatory processes and reducing the administrative burden on municipalities and planning districts. Passed by the Conservative government of Manitoba, this legislation allows the Municipal Board, who are appointees of the government, to overturn decisions made by the democratically elected officials of a municipality.

Biotechs

Shelters open to the air with straw-based manure management systems.

Concerned Citizens of Big Island – CCBI

A group of residents affected by the hog barn that has been working together since 2004 to try and deal with the situation and the consequences of the barn located so close to our community.

Conditional Use Hearing

A hearing conducted to allow people to raise any concerns about a proposed development. These concerns could include, but are not limited to, adverse effects on the amenities, use, safety, and convenience of the adjoining properties and adjacent areas.

Ducks Unlimited Canada

A not-for-profit organization dedicated to conserving, restoring and managing wetlands and associated habitats for North America's waterfowl. These habitats also benefit other wildlife and people.

Farm Practices Protection Board – FPPB

Determines what constitutes a normal farm practice for agricultural operations, and considers nuisance complaints against those operations from people directly affected by any disturbance. This has been replaced by the Manitoba Farm Industry Board.

Farrow-to-finish

A confinement operation where pigs are bred and raised to their slaughter weight, usually 225-300 lbs.

Farrow-to-weanling

Operations that oversee the breeding of herds and raise pigs until they are weaned between 10 and 15 lbs when they are sold to wean-to-finish farms.

Freedom of Information and Protection of Privacy Act

A right of access to information in records held by public bodies.

Friends of Sturgeon Creek

An environmental group based in Winnipeg concerned with the health of Sturgeon Creek and the surrounding area.

Hog Watch Manitoba

A non-profit organization that is a coalition of environmentalists, farmers, friends of animals, social justice advocates, trade unions, and scientists. They promote a hog industry in Manitoba that is ethically, environmentally, and economically sustainable.

Human Waste Lagoon

A large pond into which the sewage or effluent from a sewage system flows. Light, warmth, and oxygen helps algae and bacteria break down the sewage and effluent.

Manitoba Farm Industry Board – MFIB

Administers and enforces The Farm Lands Ownership Act, and has the authority to exempt individuals, corporations, farm land, and interest in farm land from any conditions of the legislation. This replaced the Farm Practices Protection Board.

Manitoba Water Stewardship

A department of the government of Manitoba that regulated several aspects of water stewardship in Manitoba, including water quality, fisheries, and water use licensing. The department has since been dissolved and its mandate transferred to Manitoba Conservation and Climate.

Manitoba Sustainable Development

A department of the government of Manitoba that delivered programs and services to protect the environment while sustaining and conserving Manitoba's diverse ecosystems and natural resources. Its mandate has now been transferred to Manitoba Conservation and Climate.

Manure Management Plan – MMP

A plan that helps identify when, where, and at what rate to spread manures, slurry, dirty water, and other organic wastes.

Manure Storage Facility – MSF

A storage facility for liquid manure in either open or covered watertight tanks, or lined lagoons. The lagoon is a pit dug in the soil, normally a large rectangular structure with sloping earth bank walls lined with clay.

Rural Municipality – RM

A single administrative division with corporate status and powers of self-government or jurisdiction as granted by national and regional laws to which it is subordinate.

Sow-to-weanling

Operations where pigs are sold as soon as they are weaned off their mother.

Technical Review

A review to determine whether a proposal will create a risk to health, safety, or the environment, be detrimental to the health or general welfare of people living or working in the surrounding area, and/or negatively affect other properties.

TESTIMONIALS

TO KNOW BILL MASSEY AND HIS PARTNER To know Bill Massey and his partner Dorothy Braun has been one of life's special opportunities for me. Bill's story helps us understand what it means to be in for the long haul as an active community member.

The old adage that it takes a community to raise a child can be understood as observed throughout his life. The support of his neighbours, the influence of his grandfather, and the love of his sister carried him through his early years. Also his attachment to horses and dogs has helped mold his personality. Bill is dedicated to family, friends, community and his pigs.

Solomon Cleaver, in his book "Jean Val Jean" writes, and I quote, "Isn't it wrong not to do all the good you can?" The people of Big Island and surrounding communities are fortunate to have his leadership. And from the book, a Hutterite friend told him amidst the ongoing 'boar war', "I would feel better if you were a quitter."

In his own unique way, Bill has woven his life experiences and leadership skills into the ongoing work of the Concerned Citizens of Big Island. In this book, he explains their struggle with a neighbouring community's hog barn expansion and the problems it has created, particularly with respect to animal care, and air and water contamination.

Bill and the group have fought the good fight for the better part of fifteen years, despite being constantly let down and thwarted in their efforts by government departments and individuals at all levels.

Of particular disappointment to me, as a farmer and a past NDP MLA is the complete failure of the Manitoba Ombudsman to support the issues raised and documented by the Concerned Citizens of Big Island. The Ombudsman's office investigates a community complaint about the fairness of government actions, decisions, or serious wrong-doings, and its purpose is to defend individuals and groups who have exhausted all other avenues. Unfortunately, as you will see in Bill's book, this office has developed into another layer of bureaucrats that stymie those most in need of help.

Bill Massey is no quitter, and this book shows that not only is he unafraid to fight the good fight, but also he has lived a life that allows him to do that as long as necessary.

— Walter Jess, farmer and past Saskatchewan
NDP MLA 1991-1999

OF PORK & POTATOES IS THE FIRST memoir I have edited that is actually two main stories in one. And author Bill Massey has not only managed to blend both in a delightful and telling way, but also to bridge between the past and the present until they finally converge in the last chapter. I say the 'last' chapter because I have the distinct impression that there are more chapters to come.

What got me interested in editing this book is Bill's decision not to write the two stories in chronological order. Right from the outset, he opted instead to allow each to stand on their own and to complement the other in telling the whole story.

As told primarily in the even-numbered chapters, Bill's personal story explores his life growing up and working on farms in rural Manitoba, his career as a teacher, and raising his own family while farming on the prairies. We meet his parents and siblings, his partner and wife Dorothy, his children and grandchildren, and other key individuals who have played various roles in his life.

The second story—told in the odd-numbered chapters—is about his and his community's more than 15-year struggle against business and bureaucracy at a number of levels.

Both stories work together to show how Bill's love of the land and his education and leadership skills equipped him with the necessary knowledge, passion and resilience to take on one of the biggest challenges of his 'retirement'. And each chapter in Bill's life provides clues to why he was chosen by his community to lead them in that struggle.

During the editing process, not only was Bill open to the many edits and suggestions, but he was also willing to extend the story where the telling required it. For the price of admission, I learned something about the hog industry in Manitoba and the ongoing struggles in some rural communities to co-exist with that industry.

It has been a real pleasure working with Bill on his book. His determination to tell the best story possible, and his willingness to share the ups and downs of a life well lived, is a testament to his commitment to stand up for the little guy and to fight for truth however and wherever he can.

— Jenny Gates, book editor

BILL MASSEY TAKES US THROUGH HIS LIFE journey and like a tapestry, time and space are woven together in a blend of nostalgia and reality. But this is not your garden variety rendition of "this is my life". What holds the reader's interest are its vivid recollections and life lessons. The reader is barely settled into an interesting comfortable read when with deliberate execution, Bill transitions the reader into a harsh reality where large corporations, compliant governments and money are the rule. Bill's unique ability leads us into a world of facts, figures and information that underpin the hog industry.

Of Pork & Potatoes is a discovery. The reader sees how events and people throughout a childhood and adolescence can shape a man's character and prepare him for a destiny he could never conceive and challenges he could never imagine. Years later Bill comes nose to nose with the overwhelming momentum behind a burgeoning hog industry and forces supporting it. Is there respect for citizens and small communities who want their rights recognized? Are there rights? If the reader is uncomfortable with this reality, it is because they are learning throughout reading this book how one man shaped by his past was destined to confront an industry that few of us ever give much thought or due consideration.

As a reader, I know I will never pick up a package of pork without a deeper understanding of the hog industry, or questioning those unseeingly political and money driven elements hidden from the public eye and lurking behind all that packaging.

The author has carefully crafted this book leading us gently towards sharp realizations. We are all better for reading this account. We find out more than just the story of one man's journey. If we allow our minds to be curious, our hearts to be open and our courage to be found, we can see how our personal pathways of life can lead towards an advocacy never considered.

— Maureen Mitchells, Member of the Concerned Citizens of Big Island and former journalist

ABOUT THE AUTHOR

IN RURAL MANITOBA, BILL MASSEY TAUGHT FOR ten years in elementary schools and, at the age of thirty, became a principal for twenty-three years. He was a children's advocate, piloting child abuse prevention programs and sitting on the abuse committee for twenty years, representing his school division. He became a specialist in dealing with violent children and taught Nonviolent Crisis Intervention at the University of Winnipeg and Urban Circle in Winnipeg for fifteen years. During all that time, he has lived on farms at Kelwood, Elma, and finally Grosse Isle, where he currently resides with his wonderful wife and partner, Dorothy. He enjoys history, astronomy, repairing old farm equipment and farming.

Printed in Canada